**McGraw-Hill My Math**

# Interactive Guide

Grade 5

ConnectED.mcgraw-hill.com

Copyright © 2014 McGraw-Hill Education

All rights reserved. No part of this publication may be reproduced or distributed in any form or by any means, or stored in a database or retrieval system, without the prior written consent of McGraw-Hill Education, including, but not limited to, network storage or transmission, or broadcast for distance learning.

**STEM** McGraw-Hill is committed to providing instructional materials in Science, Technology, Engineering, and Mathematics (STEM) that give all students a solid foundation, one that prepares them for college and careers in the 21st century.

Send all inquiries to:
McGraw-Hill Education
8787 Orion Place
Columbus, OH 43240

Selections from:
ISBN: 978-0-02-130831-6 *(Grade 5 Student Edition)*
MHID: 0-02-130831-4 *(Grade 5 Student Edition)*
ISBN: 978-0-02-130237-6 *(Grade 5 Teacher Edition)*
MHID: 0-02-130237-5 *(Grade 5 Teacher Edition)*

Printed in the United States of America.

Visual Kinesthetic Vocabulary® is a registered trademark of Dinah-Might Adventures, LP.

10 LOV 20 19 18

# Contents

## Chapter 1 Place Value

Mathematical Practice 1/Inquiry . . . . . 1
Lesson 1 Place Value Through Millions . . . . . 2
Lesson 2 Compare and Order Whole Numbers Through Millions . . . . . 3
Lesson 3 Inquiry/Hands On: Model Fractions and Decimals . . . . . 4
Lesson 4 Represent Decimals . . . . . 5
Lesson 5 Inquiry/Hands On: Understand Place Value . . . . . 6
Lesson 6 Place Value Through Thousandths . . . . . 7
Lesson 7 Compare Decimals . . . . . 8
Lesson 8 Order Whole Numbers and Decimals . . . . . 9
Lesson 9 Problem-Solving Investigation: Use the Four-Step Plan . . . . . 10

## Chapter 2 Multiply Whole Numbers

Mathematical Practice 3/Inquiry . . . . . 11
Lesson 1 Prime Factorization . . . . . 12
Lesson 2 Inquiry/Hands On: Prime Factorization Patterns . . . . . 13
Lesson 3 Powers and Exponents . . . . . 14
Lesson 4 Multiplication Patterns . . . . . 15
Lesson 5 Problem-Solving Investigation: Make a Table . . . . . 16
Lesson 6 Inquiry/Hands On: Use Partial Products and the Distributive Property . . . . . 17
Lesson 7 The Distributive Property . . . . . 18
Lesson 8 Estimate Products . . . . . 19
Lesson 9 Multiply by One-Digit Numbers . . . . . 20
Lesson 10 Multiply by Two-Digit Numbers . . . . . 21

## Chapter 3 Divide by a One-Digit Divisor

Mathematical Practice 8/Inquiry . . . . . 22
Lesson 1 Relate Division to Multiplication . . . . . 23
Lesson 2 Inquiry/Hands On: Division Models . . . . . 24
Lesson 3 Two-Digit Dividends . . . . . 25
Lesson 4 Division Patterns . . . . . 26
Lesson 5 Estimate Quotients . . . . . 27
Lesson 6 Inquiry/Hands On: Division Models with Greater Numbers . . . . . 28
Lesson 7 Inquiry/Hands On: Distributive Property and Partial Quotients . . . . . 29
Lesson 8 Divide Three-and Four-Digit Dividends . . . . . 30
Lesson 9 Place the First Digit . . . . . 31
Lesson 10 Quotients with Zeros . . . . . 32
Lesson 11 Inquiry/Hands On: Use Models to Interpret the Remainder . . . . . 33
Lesson 12 Interpret the Remainder . . . . . 34
Lesson 13 Problem-Solving Investigation: Extra or Missing Information . . . . . 35

## Chapter 4 Divide by a Two-Digit Divisor

Mathematical Practice 8/Inquiry . . . . . 36
Lesson 1 Estimate Quotients . . . . . 37
Lesson 2 Inquiry/Hands On: Divide Using Base-Ten Blocks . . . . . 38
Lesson 3 Divide by a Two-Digit Divisor . . . . . 39
Lesson 4 Adjust Quotients . . . . . 40
Lesson 5 Divide Greater Numbers . . . . . 41
Lesson 6 Problem-Solving Investigation: Solve a Simpler Problem . . . . . 42

## Chapter 5 Add and Subtract Decimals

| | |
|---|---|
| Mathematical Practice 2/Inquiry | 43 |
| Lesson 1 Round Decimals | 44 |
| Lesson 2 Estimate Sums and Differences | 45 |
| Lesson 3 Problem-Solving Investigation: Estimate or Exact Answer | 46 |
| Lesson 4 Inquiry/Hands On: Add Decimals Using Base-Ten Blocks | 47 |
| Lesson 5 Solve Hands On: Add Decimals Using Models | 48 |
| Lesson 6 Add Decimals | 49 |
| Lesson 7 Addition Properties | 50 |
| Lesson 8 Inquiry/Hands On: Subtract Decimals Using Base-Ten Blocks | 51 |
| Lesson 9 Inquiry/Hands On: Subtract Decimals Using Models | 52 |
| Lesson 10 Subtract Decimals | 53 |

## Chapter 6 Multiply and Divide Decimals

| | |
|---|---|
| Mathematical Practice 2/Inquiry | 54 |
| Lesson 1 Estimate Products of Whole Numbers and Decimals | 55 |
| Lesson 2 Inquiry/Hands On: Use Models to Multiply | 56 |
| Lesson 3 Multiply Decimals by Whole Numbers | 57 |
| Lesson 4 Inquiry/Hands On: Use Models to Multiply Decimals | 58 |
| Lesson 5 Multiply Decimals | 59 |
| Lesson 6 Multiply Decimals by Powers of Ten | 60 |
| Lesson 7 Problem-Solving Investigation: Look for a Pattern | 61 |
| Lesson 8 Multiplication Properties | 62 |
| Lesson 9 Estimate Quotients | 63 |
| Lesson 10 Inquiry/Hands On: Divide Decimals | 64 |
| Lesson 11 Divide Decimals by Whole Numbers | 65 |
| Lesson 12 Inquiry/Hands On: Use Models to Divide Decimals | 66 |
| Lesson 13 Divide Decimals | 67 |
| Lesson 14 Divide Decimals by Powers of Ten | 68 |

## Chapter 7 Expressions and Patterns

| | |
|---|---|
| Mathematical Practice 4/Inquiry | 69 |
| Lesson 1 Inquiry/Hands On: Numerical Expressions | 70 |
| Lesson 2 Order of Operations | 71 |
| Lesson 3 Write Numerical Expressions | 72 |
| Lesson 4 Problem-Solving Investigation: Work Backward | 73 |
| Lesson 5 Inquiry/Hands On: Generate Patterns | 74 |
| Lesson 6 Patterns | 75 |
| Lesson 7 Inquiry/Hands On: Map Locations | 76 |
| Lesson 8 Ordered Pairs | 77 |
| Lesson 9 Graph Patterns | 78 |

## Chapter 8 Fractions and Decimals

| | |
|---|---|
| Mathematical Practice 3/Inquiry | 79 |
| Lesson 1 Fractions and Division | 80 |
| Lesson 2 Greatest Common Factor | 81 |
| Lesson 3 Simplest Form | 82 |
| Lesson 4 Problem-Solving Investigation: Guess, Check, and Revise | 83 |
| Lesson 5 Least Common Multiple | 84 |
| Lesson 6 Compare Fractions | 85 |
| Lesson 7 Inquiry/Hands On: Use Models to Write Fractions as Decimals | 86 |
| Lesson 8 Write Fractions as Decimals | 87 |

## Chapter 9 Add and Subtract Fractions

| | |
|---|---|
| Mathematical Practice 5/Inquiry | 88 |
| Lesson 1 Round Fractions | 89 |
| Lesson 2 Add Like Fractions | 90 |
| Lesson 3 Subtract Like Fractions | 91 |
| Lesson 4 Inquiry/Hands On: Use Models to Subtract Unlike Fractions | 92 |
| Lesson 5 Add Unlike Fractions | 93 |
| Lesson 6 Inquiry/Hands On: Use Models to Subtract Unlike Fractions | 94 |
| Lesson 7 Subtract Unlike Fractions | 95 |
| Lesson 8 Problem-Solving Investigation: Determine Reasonable Answers | 96 |
| Lesson 9: Estimate Sums and Differences | 97 |
| Lesson 10 Inquiry/Hands On: Use Models to Add Mixed Numbers | 98 |
| Lesson 11 Add Mixed Numbers | 99 |
| Lesson 12 Subtract Mixed Numbers | 100 |
| Lesson 13 Subtract with Renaming | 101 |

## Chapter 10 Multiply and Divide Fractions

| | |
|---|---|
| Mathematical Practice 5/Inquiry | 102 |
| Lesson 1 Inquiry/Hands On: Part of a Number | 103 |
| Lesson 2 Estimate Products of Fractions | 104 |
| Lesson 3 Inquiry/Hands On: Model Fraction Multiplication | 105 |
| Lesson 4 Multiply Whole Numbers and Fractions | 106 |
| Lesson 5 Inquiry/Hands On: Use Models to Multiply Fractions | 107 |
| Lesson 6 Multiply Fractions | 108 |
| Lesson 7 Multiply Mixed Numbers | 109 |
| Lesson 8 Inquiry/Hands On: Multiplication as Scaling | 110 |
| Lesson 9 Inquiry/Hands On: Division with Unit Fractions | 111 |
| Lesson 10 Divide Whole Numbers by Unit Fractions | 112 |
| Lesson 11 Divide Unit Fractions by Whole Numbers | 113 |
| Lesson 12 Problem-Solving Investigation: Draw a Diagram | 114 |

## Chapter 11 Measurement

Mathematical Practice 6/Inquiry . . . 115

Lesson 1 Inquiry/Hands On: Measure with a Ruler . . . 116

Lesson 2 Convert Customary Units of Length . . . 117

Lesson 3 Problem-Solving Investigation: Use Logical Reasoning . . . 118

Lesson 4 Inquiry/Hands On: Estimate and Measure Weight . . . 119

Lesson 5 Convert Customary Units of Weight . . . 120

Lesson 6 Inquiry/Hands On: Estimate and Measure Capacity . . . 121

Lesson 7 Convert Customary Units of Capacity . . . 122

Lesson 8 Display Measurement Data on a Line Plot . . . 123

Lesson 9 Inquiry/Hands On: Metric Rulers . . . 124

Lesson 10 Convert Metric Units of Length . . . 125

Lesson 11 Inquiry/Hands On: Estimate and Measure Metric Mass . . . 126

Lesson 12 Convert Metric Units of Mass . . . 127

Lesson 13 Convert Metric Units of Capacity . . . 128

## Chapter 12 Geometry

Mathematical Practice 7/Inquiry . . . 129

Lesson 1 Polygons . . . 130

Lesson 2 Inquiry/Hands On: Sides and Angles of Triangles . . . 131

Lesson 3 Classify Triangles . . . 132

Lesson 4 Inquiry/Hands On: Sides and Angles of Quadrilaterals . . . 133

Lesson 5 Classify Quadrilaterals . . . 134

Lesson 6 Inquiry/Hands On: Build Three-Dimensional Figures . . . 135

Lesson 7 Three-Dimensional Figures . . . 136

Lesson 8 Inquiry/Hands On: Use Models to Find Volume . . . 137

Lesson 9 Volume of Prisms . . . 138

Lesson 10 Inquiry/Hands On: Build Composite Figures . . . 139

Lesson 11 Volume of Composite Figures . . . 140

Lesson 12 Problem-Solving Investigation: Make a Model . . . 141

Visual Kinesthetic Vocabulary© . . . VKV1

# Chapter 1 Place Value

## Inquiry of the Essential Question:

**How does the position of a digit in a number relate to its value?**

Read the Essential Question. Describe your observations (I see...), inferences (I think...), and prior knowledge (I know...) of each math example. Write additional questions you have below. Then share your ideas and questions with a classmate.

| Thousands Period | | | Ones Period | | |
|---|---|---|---|---|---|
| hundreds | tens | ones | hundreds | tens | ones |
| 2 | 7 | 7 | 3 | 8 | 9 |

I see …

I think…

I know…

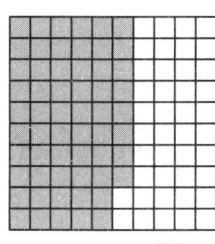

Fraction: $\frac{58}{100}$

Word form: fifty-eight hundredths

Decimal: 0.58

I see …

I think…

I know…

**Step 1** Understand the facts, and what needs to be found.
**Step 2** Plan the strategy.
**Step 3** Solve the problem.
**Step 4** Check that the answer makes sense.

I see …

I think…

I know…

Questions I have…

_____

_____

_____

NAME _____ DATE _____

# Lesson 1 Vocabulary Chart

*Place Value Through Millions*

Use the three-column chart to organize the vocabulary in this lesson. Write the word in Spanish. Then write the correct terms to complete each definition.

| English | Spanish | Definition |
|---|---|---|
| **period** | | Each group of _____ digits on a place-value chart. |
| **standard form** | | The usual or common way to write a number using _____. |
| **expanded form** | | A way of writing a number as the _____ of the _____ of its digits. |
| **place** | | The _____ of a digit in a _____. |
| **place value** | | The value given to a _____ by its _____ in a number. |
| **place-value chart** | | A chart that shows the _____ of the _____ in a number. |

2 Grade 5 • Chapter 1 *Place Value*

# Lesson 2 Concept Web

*Compare and Order Whole Numbers through Millions*

Use the concept web to identify which symbol to use, greater than (>), less than (<), or equal to (=), for each example.

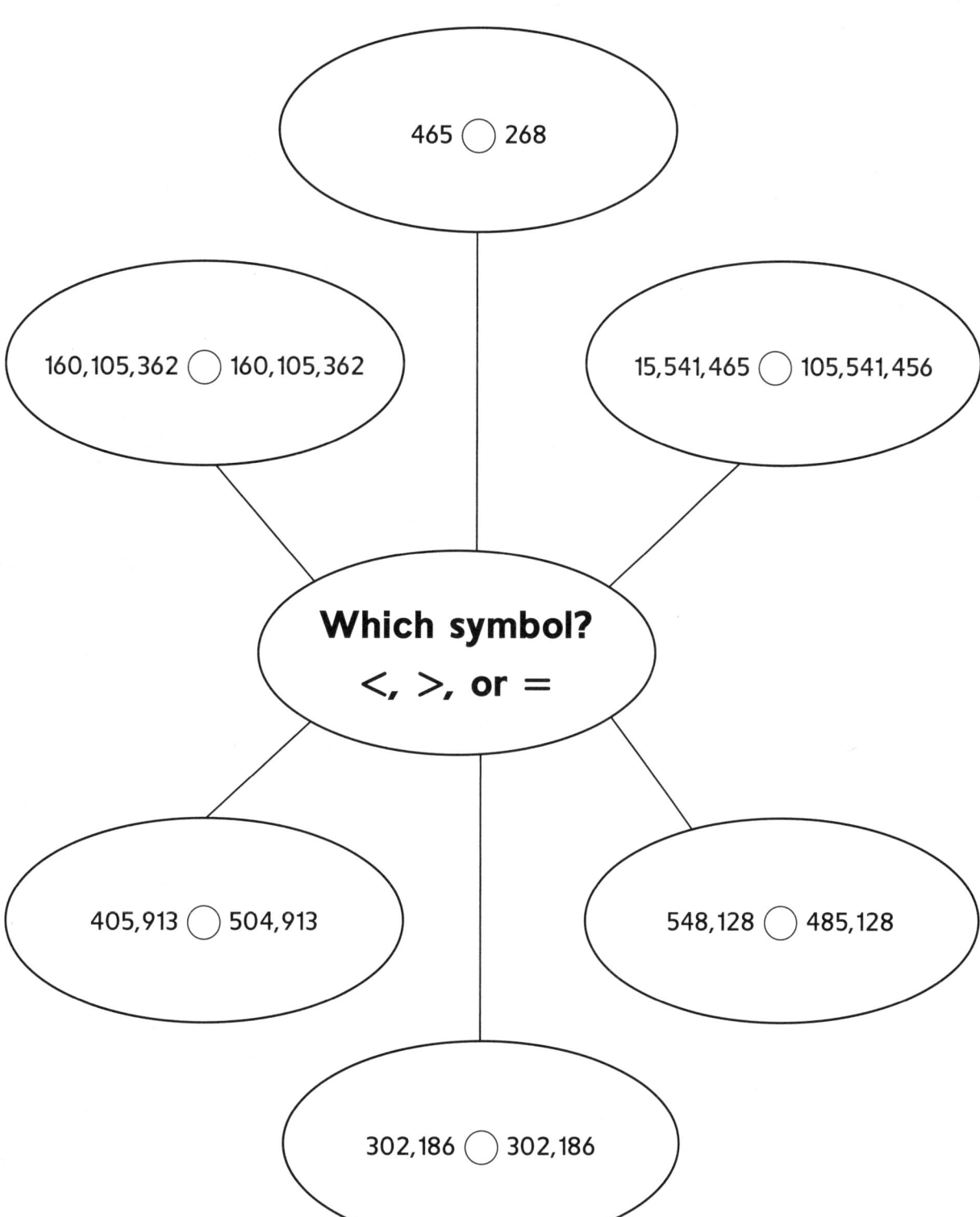

Grade 5 • Chapter 1 *Place Value*  3

NAME _____ DATE _____

# Lesson 3 Vocabulary Cognates

*Inquiry/Hands On: Model Fractions and Decimals*

Use the Glossary to define the math word in English and in Spanish in the word boxes. Write a sentence using your math word.

| **decimal** | **decimal** |
|---|---|
| Definition | Definición |
| My math word sentence: | |

| **decimal point** | **punto decimal** |
|---|---|
| Definition | Definición |
| My math word sentence: | |

**4** Grade 5 • Chapter 1 *Place Value*

NAME _____ DATE _____

# Lesson 4 Vocabulary Definition Map
*Represent Decimals*

Use the definition map to write a description and list characteristics about the vocabulary word or phrase. Write or draw math examples. Share your examples with a classmate.

My Math Vocabulary:

**decimal**

Characteristics from Lesson:

A number written as a decimal contains a decimal point which is a period between the _____ and the _____ place.

A decimal in word form is similar to the word form of whole numbers, but it contains the ending - ____.

10: ten

$\frac{1}{10}$ or 0.1: tenths

100: hundred

$\frac{1}{100}$ or 0.01: _____

1,000: thousand

$\frac{1}{1,000}$ or 0.001: _____

Description from Glossary:

My Math Example:

Model a decimal by shading _____ squares that make up a _____ square. For example, shade 32 of the 100 squares to model the decimal _____.

Grade 5 • Chapter 1 *Place Value* 5

NAME _____  DATE _____

# Lesson 5 Guided Writing

## Inquiry/Hands On: Understand Place Value

**How do you use place value to understand decimals?**

Use the exercises below to help you build on answering the Essential Question. Write the correct word or phrase on the lines provided.

1. Rewrite the question in your own words.

   _____

   _____

2. What key words do you see in the question?

   _____

3. Place value is the value given to a _____ by its _____ in a number.

4. The value of the digit 2 in the **ones** place is ___ or _____ ones.
   The value of the digit 2 in the **tens** place is ___ or _____ tens.

   | tens | ones |
   |------|------|
   | 2    | 2    |

5. The value of the digit 2 in the **tens** place is ___ times as much as the value of the digit 2 in the ones place. The value of the digit 2 in the **ones** place is $\frac{\square}{\square}$ times as much as the value of the digit 2 in the tens place.

6. The value of the digit 2 in the **tenths** place is _____ or _____ tenths.
   The value of the digit 2 in the **hundredths** place is _____ or _____ hundredths.

   | ones | tenths | hundredths |
   |------|--------|------------|
   | 0    | 2      | 2          |

7. The value of the digit 2 in the **tenths** place is ___ times as much as the value of the digit 2 in the hundredths place. The value of the digit 2 in the **hundredths** place is $\frac{\square}{\square}$ times as much as the value of the digit 2 in the tenths place.

8. How do you use place value to understand decimals?

   _____

   _____

# Lesson 6 Vocabulary Chart

*Place Value Through Thousandths*

Use the concept web to identify the place of each digit in the decimal. Write in word form.

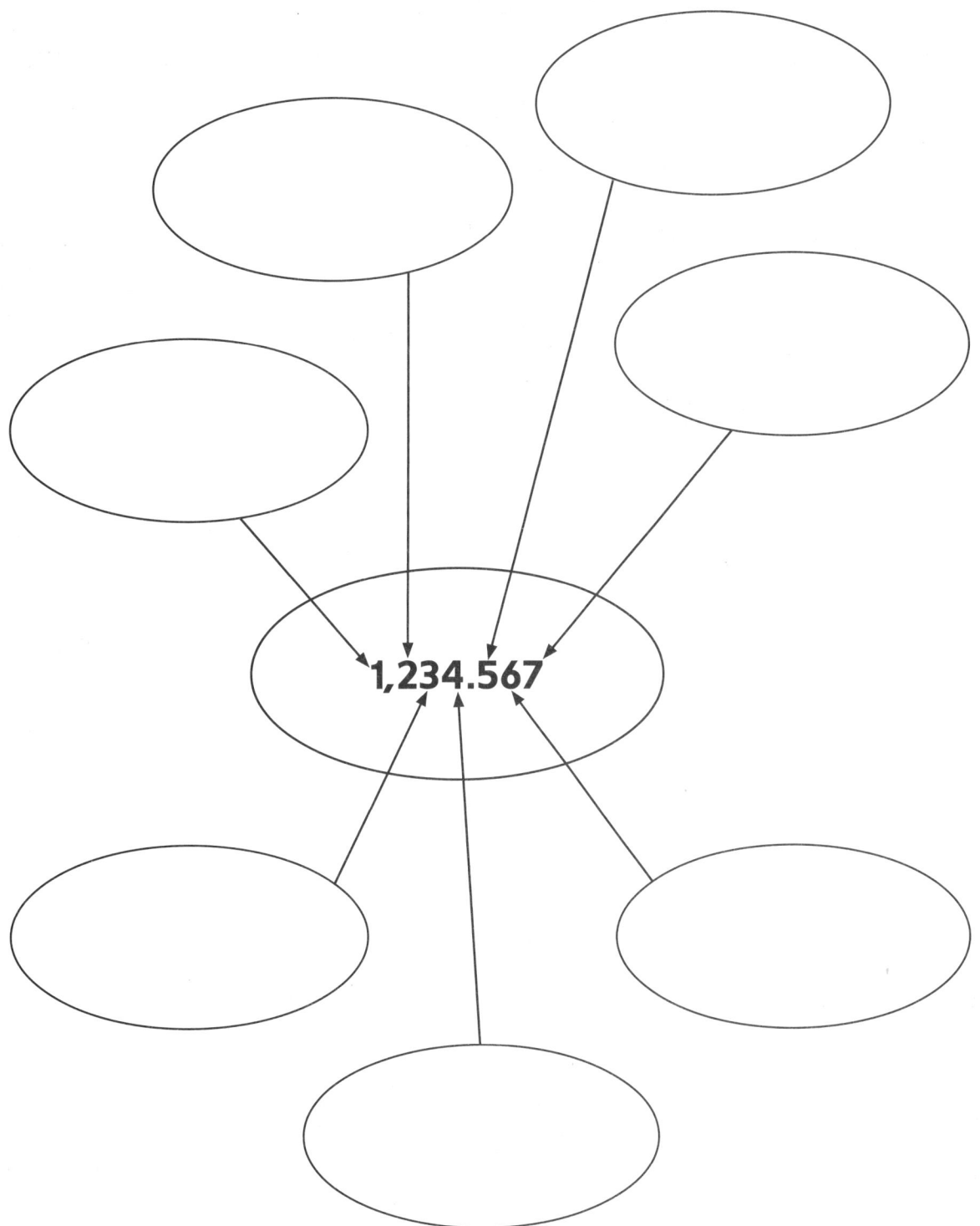

Grade 5 • Chapter 1 *Place Value* 7

NAME _____  DATE _____

# Lesson 7 Four-Square Vocabulary

*Compare Decimals*

Write the definition for each math word. Write what each word means in your own words. Draw or write examples that show each math word meaning. Then write your own sentences using the words.

| Definition | My Own Words |
|---|---|
| | |
| **decimal** | |
| My Examples | My Sentence |
| | |

| Definition | My Own Words |
|---|---|
| | |
| **equivalent decimals** | |
| My Examples | My Sentence |
| | |

NAME _____ DATE _____

# Lesson 8 Note Taking

## Order Whole Numbers and Decimals

Read the question. Write words you need help with and research each word. Use your lesson to write your Cornell notes. Write or draw math examples to explain your thinking. Share your examples with a classmate.

| Building on the Essential Question | Notes: |
|---|---|
| How do you order whole numbers and decimals? | When comparing numbers, start comparing the numbers using the _____ place value. |
| | Place value is the value given to a digit by its _____ in a number. |
| | When you look at the place values of a decimal, each digit to the left has a value that is _____ times as much as that same digit would have in the place to its right. |
| | The hundreds place is _____ than the tens place. |
| | The **ten*ths*** place is _____ than the **hundred*ths*** place. |
| Words I need help with: | You can also compare numbers by locating the numbers on a number line. A number line is a _____ that represents numbers as _____. |
| | When you look at a number line, each decimal number to the right of another decimal is _____ than that number. |
| | After locating two decimals on a number line, the decimal to the _____ is greater. |

**My Math Examples:**

Grade 5 • Chapter 1 *Place Value* 9

NAME _____ DATE _____

# Lesson 9 Problem-Solving Investigation

## STRATEGY: Use the Four-Step Plan

Use the four-step plan to solve each problem.

1. The **table** shows the number of **ounces** of butter **Marti** used in different recipes.
   She (Marti) has **6 ounces** of butter left.
   How many **ounces** of butter did **she** have at the **beginning**?

| Recipe | Ounces of Butter |
|---|---|
| Pie | 4 |
| Cookie | 8 |
| Pasta | 6 |

| Understand | Solve |
|---|---|
| I know: <br><br> I need to find: | |
| **Plan** | **Check** |
| | |

2. At the end of their **3-day** vacation, the **Palmers** traveled a **total** of **530** miles.
   On the **third** day, they drove **75** miles.
   On the **second** day, they drove **320** miles.
   How many miles did **they** (the Palmers) drive the **first** day?

| Understand | Solve |
|---|---|
| I know: <br><br> I need to find: | |
| **Plan** | **Check** |

| Day | Miles Driven |
|---|---|
| 1 | |
| 2 | |
| 3 | |

10 Grade 5 • Chapter 1 *Place Value*

# Chapter 2 Multiply Whole Numbers

*Inquiry of the Essential Question:*

**What strategies can be used to multiply whole numbers?**

Read the Essential Question. Describe your observations (I see..), inferences (I think...), and prior knowledge ( I know...) of each math example. Write additional questions you have below. Then share your ideas and questions with a classmate.

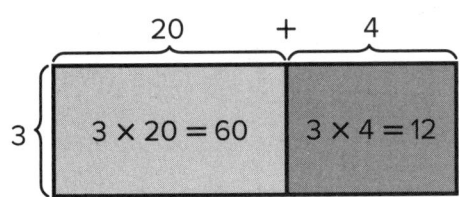

$3 \times 24 = (3 \times 20) + (3 \times 4)$
$\phantom{3 \times 24} = 60 + 12$
$\phantom{3 \times 24} = 72$

Find partial products.
Multiply.
Add.

I see ...

I think...

I know...

---

$21 \times 10^3 = 21 \times 1,000$ ← The power of 10 has three zeros.

$\phantom{21 \times 10^3} = 21,000$ ← The product has three zeros.

I see ...

I think...

I know...

---

416 — rounds to → 400
× 28 — rounds to → × 30
　　　　　　　　　12,000

I see ...

I think...

I know...

---

Questions I have...

_____

_____

_____

Grade 5 • Chapter 2 *Multiply Whole Numbers* 11

NAME _____ DATE _____

# Lesson 1 Vocabulary Definition Map
*Prime Factorization*

Use the definition map to write a description and list characteristics about the vocabulary word or phrase. Write or draw math examples. Share your examples with a classmate.

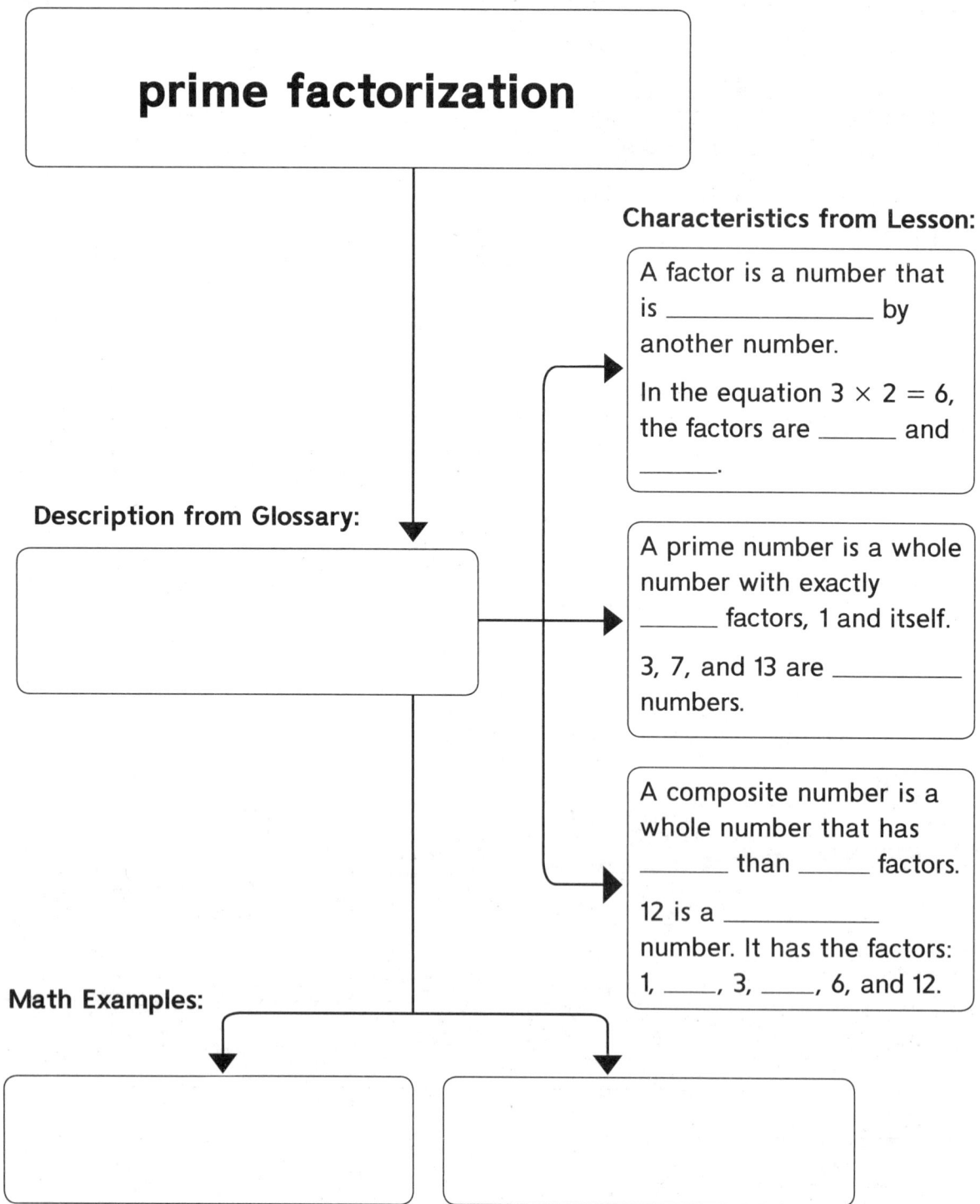

My Math Vocabulary:

**prime factorization**

**Description from Glossary:**

**Characteristics from Lesson:**

A factor is a number that is _____ by another number.

In the equation 3 × 2 = 6, the factors are _____ and _____.

A prime number is a whole number with exactly _____ factors, 1 and itself.

3, 7, and 13 are _____ numbers.

A composite number is a whole number that has _____ than _____ factors.

12 is a _____ number. It has the factors: 1, _____, 3, _____, 6, and 12.

**My Math Examples:**

12 Grade 5 • Chapter 2 *Multiply Whole Numbers*

NAME _____  DATE _____

# Lesson 2 Note Taking

## Inquiry/Hands On: Prime Factorization Patterns

Read the question. Write words you need help with and research each word. Use your lesson to write your Cornell notes. Write or draw math examples to explain your thinking. Share your examples with a classmate.

**Building on the Essential Question**

How can you find patterns in prime factorization?

**Words I need help with:**

**Notes:**

A _____ number is a whole number with exactly _____ factors, 1 and itself.

A _____ number is a whole number that has more than _____ factors.

The numbers 2 and 3 are _____ numbers.

The number 6 is a _____ number.

Prime factorization is a way of expressing a _____ number as a product of its _____ factors.

The prime factorization of _____ is 3 × 2.

The prime factorization of _____ is 3 × 2 × 2.

The prime factorization of _____ is 3 × 2 × 2 × 2.

The pattern I see is: _____
_____
_____.

**My Math Examples:**

Grade 5 • Chapter 2 Multiply Whole Numbers

NAME _____ DATE _____

# Lesson 3 Vocabulary Chart
*Powers and Exponents*

Use the three-column chart to organize the vocabulary in this lesson. Write the word in Spanish. Then write the correct terms to complete each definition.

| English | Spanish | Definition |
|---|---|---|
| **base** | | In a _____, the number used as a _____. In $10^3$, the base is 10. |
| **cubed** | | A number raised to the _____ power; $10 \times 10 \times 10$, or $10^3$. |
| **exponent** | | In a power, the number of times the _____ is used as a _____. In $5^3$, the exponent is 3. |
| **power** | | A number obtained by raising a _____ to an _____. |
| **squared** | | A number raised to the _____ power; $3 \times 3$, or $3^2$. |

14 Grade 5 • Chapter 2 *Multiply Whole Numbers*

NAME _____  DATE _____

# Lesson 4 Concept Web

## Multiplication Patterns

Use the concept web to write powers of 10 without exponents and the products of powers of 10.

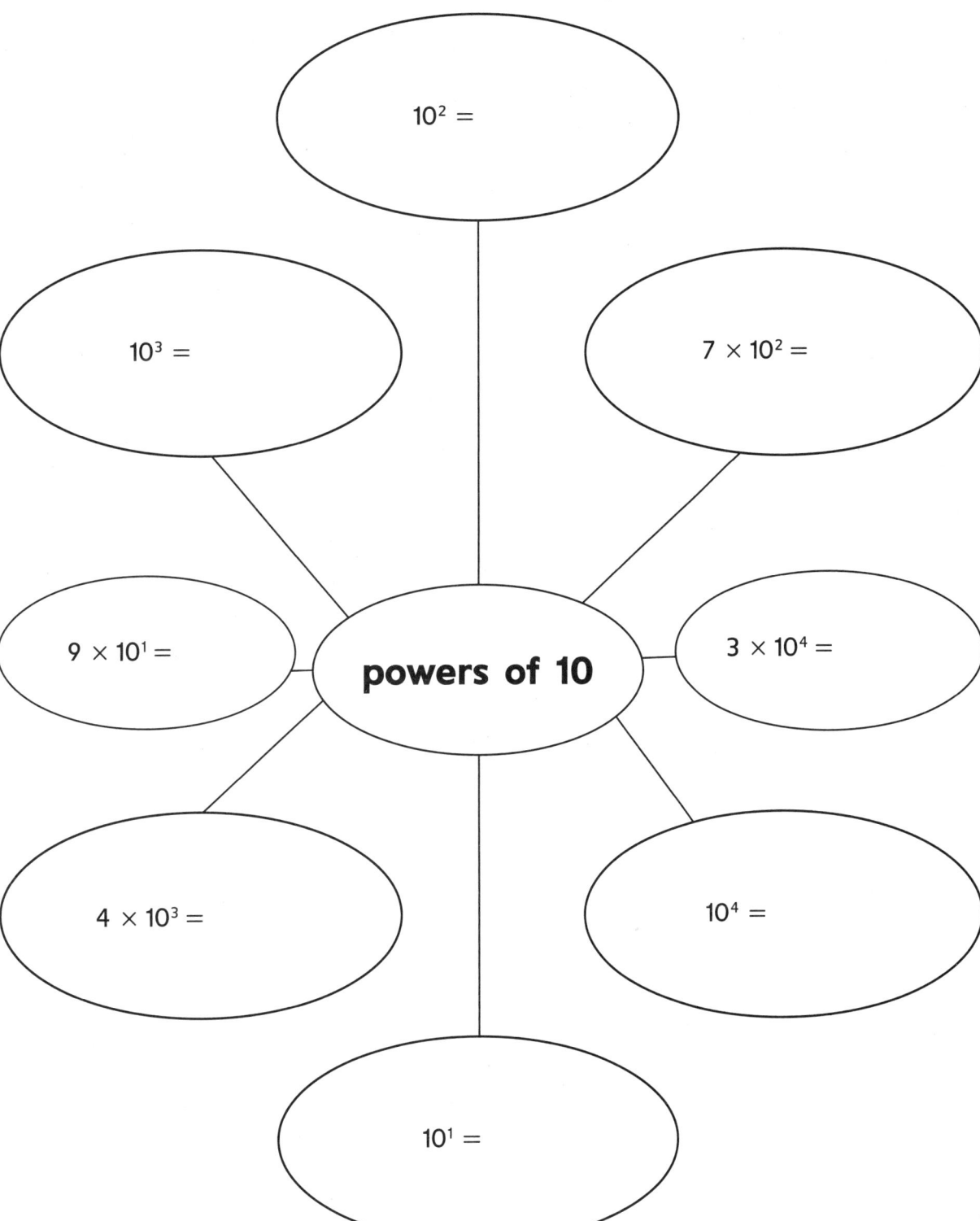

- $10^2 =$
- $10^3 =$
- $7 \times 10^2 =$
- $9 \times 10^1 =$
- powers of 10
- $3 \times 10^4 =$
- $4 \times 10^3 =$
- $10^4 =$
- $10^1 =$

Grade 5 • Chapter 2 Multiply Whole Numbers  **15**

NAME _____  DATE _____

# Lesson 5 Problem-Solving Investigation

*STRATEGY:* Make a Table

Make a table to solve each problem.

1. **Betsy** is **saving** to buy a bird cage.
   **She** (Betsy) saves **$1** the <u>first</u> week, **$3** the <u>second</u> week, **$9** the <u>third</u> week, and so on.
   How much **money** will she save in **5** weeks?

| Understand | Solve |
|---|---|
| I know: | |
| I need to find: | |

| Plan | Check |
|---|---|
| Week \| Money Saved<br>1 \| $1<br>2 \| $3<br>3 \| $5<br>4 \|<br>5 \| | |

2. **Kendall** is planning to buy a laptop for **$1,200**.
   Each month **she** <u>doubles</u> the amount she saved the <u>previous</u> month.
   If she saves **$20** the <u>first</u> month, in **how many** <u>months</u> will Kendall have **enough money** to buy the laptop?

| Understand | Solve |
|---|---|
| I know: | |
| I need to find: | |

| Plan | Check |
|---|---|
| I will make a _____. | |

16  Grade 5 • Chapter 2 *Multiply Whole Numbers*

# Lesson 6 Guided Writing

*Inquiry/Hands On: Use Partial Products and the Distributive Property*

**How do you use partial products and the Distributive Property to multiply?**

Use the exercises below to help you build on answering the Essential Question. Write the correct word or phrase on the lines provided.

1. Rewrite the question in your own words.
   _____
   _____

2. What key words do you see in the question?
   _____

3. Decompose 15 into the sum of the tens and ones.
   15 = ____ + ____

4. Rewrite the multiplication expression.
   8 × 15 = ____ × (____ + ____)

5. The _____ _____ says that to multiply a sum by a number, you can multiply each addend by the same number and add the products.

6. Use the Distributive Property to rewrite the multiplication expression.
   8 × 15 = ____ × (____ + ____) = (____ × ____) + (____ × ____)

7. Find the product of 8 × 15.
   _____

8. How do you use partial products and the Distributive Property to multiply?
   _____
   _____
   _____

**Grade 5 • Chapter 2** *Multiply Whole Numbers*   **17**

NAME _____  DATE _____

# Lesson 7 Vocabulary Definition Map

*The Distributive Property*

Use the definition map to write a description and list characteristics about the vocabulary word or phrase. Write or draw math examples. Share your examples with a classmate.

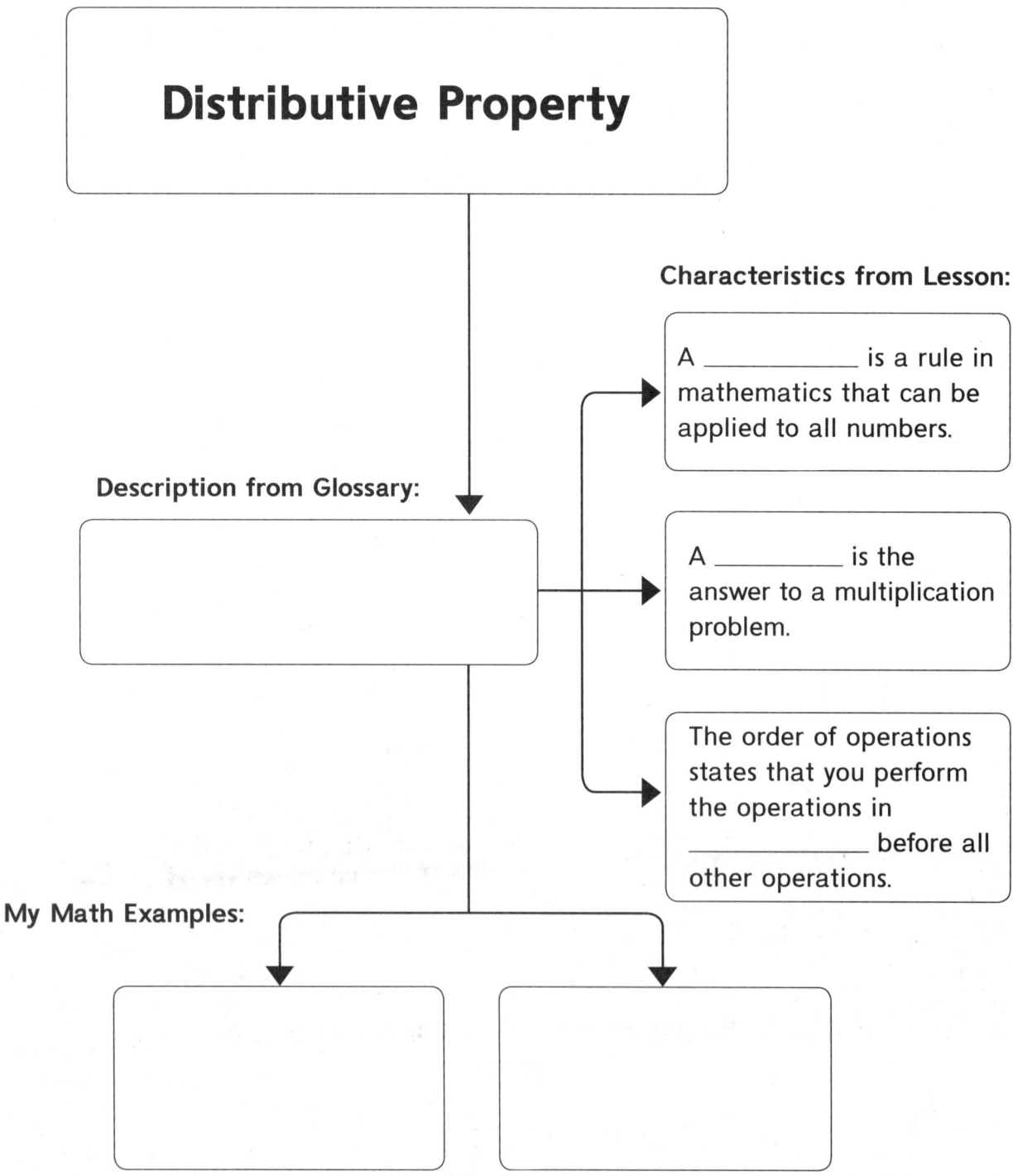

18  Grade 5 • Chapter 2 *Multiply Whole Numbers*

NAME _____ DATE _____

# Lesson 8 Note Taking

## Estimate Products

Read the question. Write words you need help with and research each word. Use your lesson to write your Cornell notes. Write or draw math examples to explain your thinking. Share your examples with a classmate.

| Building on the Essential Question | Notes: |
|---|---|
| How do you estimate products? | To round a number, find the _____ value of the number. |
| | The symbol (=) means "is _____ to." |
| | The symbol (≈) means "is _____ _____ to." |
| | Round to the nearest ten.<br>106 ≈ _____<br>47 ≈ _____ |
| | A _____ is the answer to a multiplication problem. |
| | An estimate is a number close to an _____ value. An estimate indicates _____ how much. |
| **Words I need help with:** | Estimate the product by rounding each factor to the nearest ten.<br>106 × 47 ≈ _____ × _____ = _____ |
| | _____ _____ are numbers in a problem that are easy to compute mentally. |
| | Even though 106 rounds to _____, it is easier to mentally multiply by 100. |
| | Estimate the product using compatible numbers.<br>86 × 23 ≈ _____ × _____ = _____ |
| **My Math Examples:** | |

Grade 5 • Chapter 2 Multiply Whole Numbers  **19**

NAME _____  DATE _____

# Lesson 9 Multiple Meaning Word
*Multiply by One-Digit Numbers*

Complete the four-square chart to review the multiple meaning word or phrase.

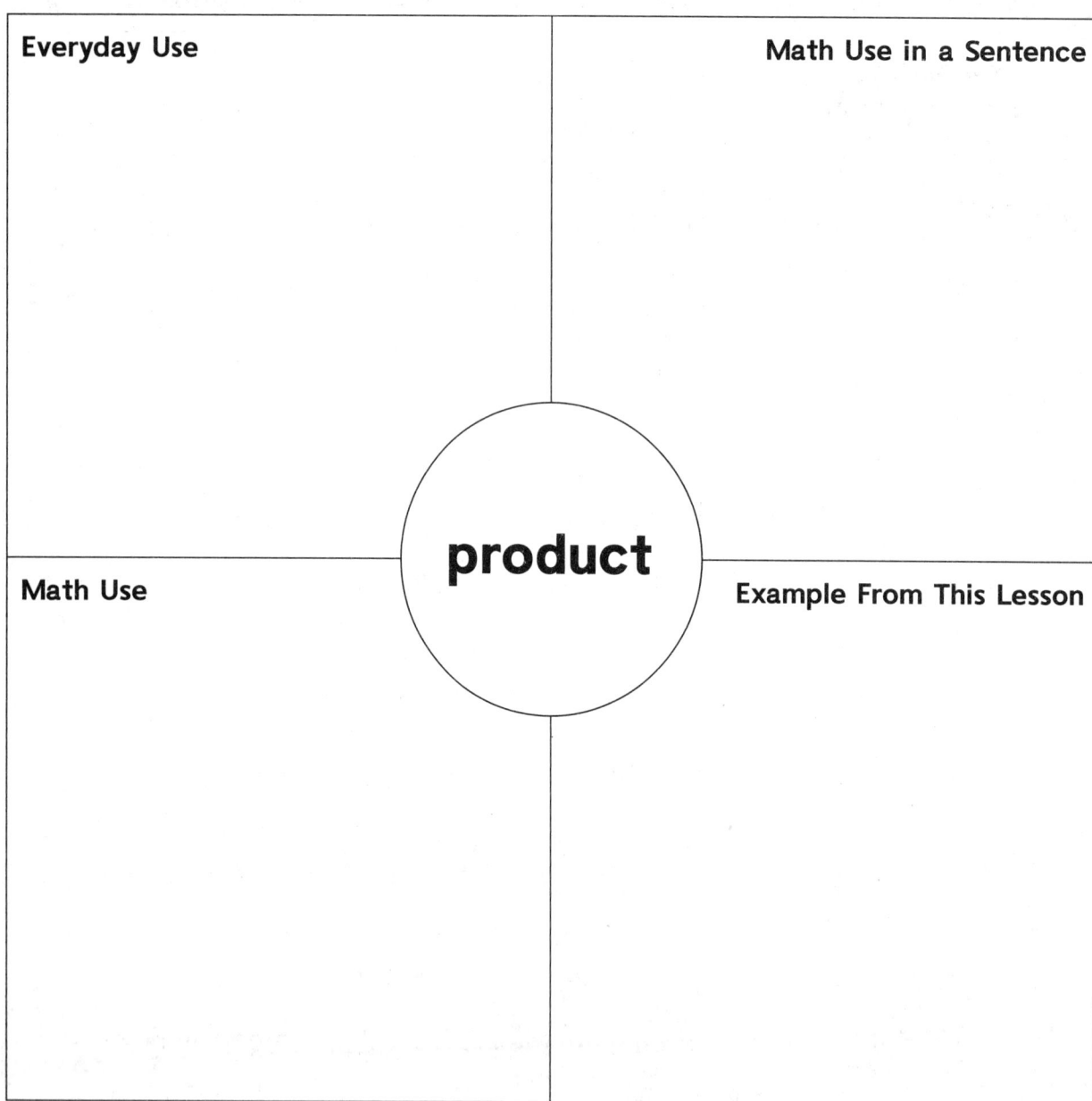

Write the correct term on each line to complete the sentence.

When using partial products to multiply, _____ the partial products to find the _____ of the _____ expression.

20  Grade 5 • Chapter 2 *Multiply Whole Numbers*

NAME _____  DATE _____

# Lesson 10 Vocabulary Cognates

*Multiply by Two-Digit Numbers*

Use the Glossary to define the math word in English and in Spanish in the word boxes. Write a sentence using your math word.

| product | producto |
|---|---|
| Definition | Definición |
| My math word sentence: | |

| estimate | estimación |
|---|---|
| Definition | Definición |
| My math word sentence: | |

Grade 5 • Chapter 2 *Multiply Whole Numbers*

NAME _____  DATE _____

# Chapter 3 Divide by a One-Digit Divisor

*Inquiry of the Essential Question:*

**What strategies can be used to divide whole numbers?**

Read the Essential Question. Describe your observations (I see...), inferences (I think...), and prior knowledge (I know...) of each math example. Write additional questions you have below. Then share your ideas and questions with a classmate.

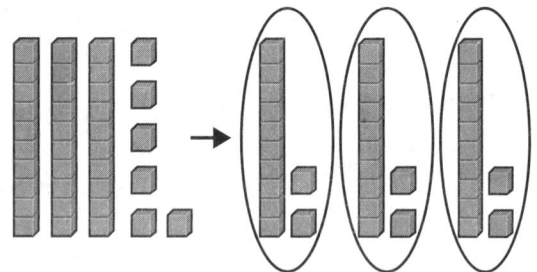

36 blocks can be arranged into 3 groups of 12 blocks each. So, 36 ÷ 3 = 12.

I see ...

I think...

I know...

|   | 80 | 4 | 1 |
|---|---|---|---|
| 5 | 400 | 20 | 5 |

$425 \div 5 = (400 + 20 + 5) \div 5$
$\phantom{425 \div 5 } = 80 + 4 + 1$
$\phantom{425 \div 5 } = 85$

I see ...

I think...

I know...

```
    128
  4)512
   -4
    11
    -8
    32
   -32
     0
```

Step 1: Divide the hundreds.

Step 2: Divide the tens.

Step 3: Divide the ones.

I see ...

I think...

I know...

Questions I have...

_____

_____

_____

NAME _____  DATE _____

# Lesson 1 Vocabulary Cognates
*Relate Division to Multiplication*

Use the Glossary to define the math word in English and in Spanish in the word boxes. Write a sentence using your math word.

| unknown | incógnita |
|---|---|
| Definition | Definición |
| My math word sentence: | |

| variable | variable |
|---|---|
| Definition | Definición |
| My math word sentence: | |

| fact family | familia de operaciones |
|---|---|
| Definition | Definición |
| My math word sentence: | |

NAME _____ DATE _____

# Lesson 2 Guided Writing

*Inquiry/Hands On: Division Models*

**How do you model division?**

Use the exercises below to help you build on answering the Essential Question. Write the correct word or phrase on the lines provided.

1. Rewrite the question in your own words.
   _____
   _____

2. What key words do you see in the question?
   _____

3. Identify the number modeled with the base-ten blocks. ____

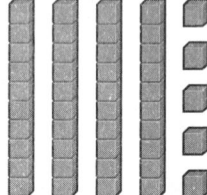

4. If you divide the tens into 3 equal groups, there will be ____ ten in each group.

5. After dividing the tens into 3 equal groups, there will be ____ ten remaining. If you regroup the remaining ____ ten into ones, you will have ____ ones altogether.

6. If you divide the 15 ones into 3 equal groups, there will be ____ ones in each group.

7. Each group of tens and ones now has ____ ten and ____ ones.

8. So, 45 ÷ 3 = ____.

9. How do you model division?
   _____
   _____
   _____
   _____

24  Grade 5 • Chapter 3 *Divide by a One-Digit Divisor*

# Lesson 3 Vocabulary Chart

*Two-Digit Dividends*

Use the three-column chart to organize the vocabulary in this lesson. Write the word in Spanish. Then write the correct terms to complete each definition.

| English | Spanish | Definition |
|---|---|---|
| dividend | | A number that is being _____. |
| divisor | | The number that _____ the _____. <br><br> The divisor tells you how many _____. |
| quotient | | The result of a _____ problem. |
| remainder | | The number that is _____ after one whole number is _____ by another. |
| divisible or divide | | Describes a number that can be divided into _____ parts and has _____ remainder. <br><br> or <br><br> An operation on two numbers in which the first number is _____ into the same number of _____ groups as the second number. |

Grade 5 • Chapter 3 Divide by a One-Digit Divisor

NAME _____ DATE _____

# Lesson 4 Multiple Meaning Word
*Division Patterns*

Complete the four-square chart to review the multiple meaning word.

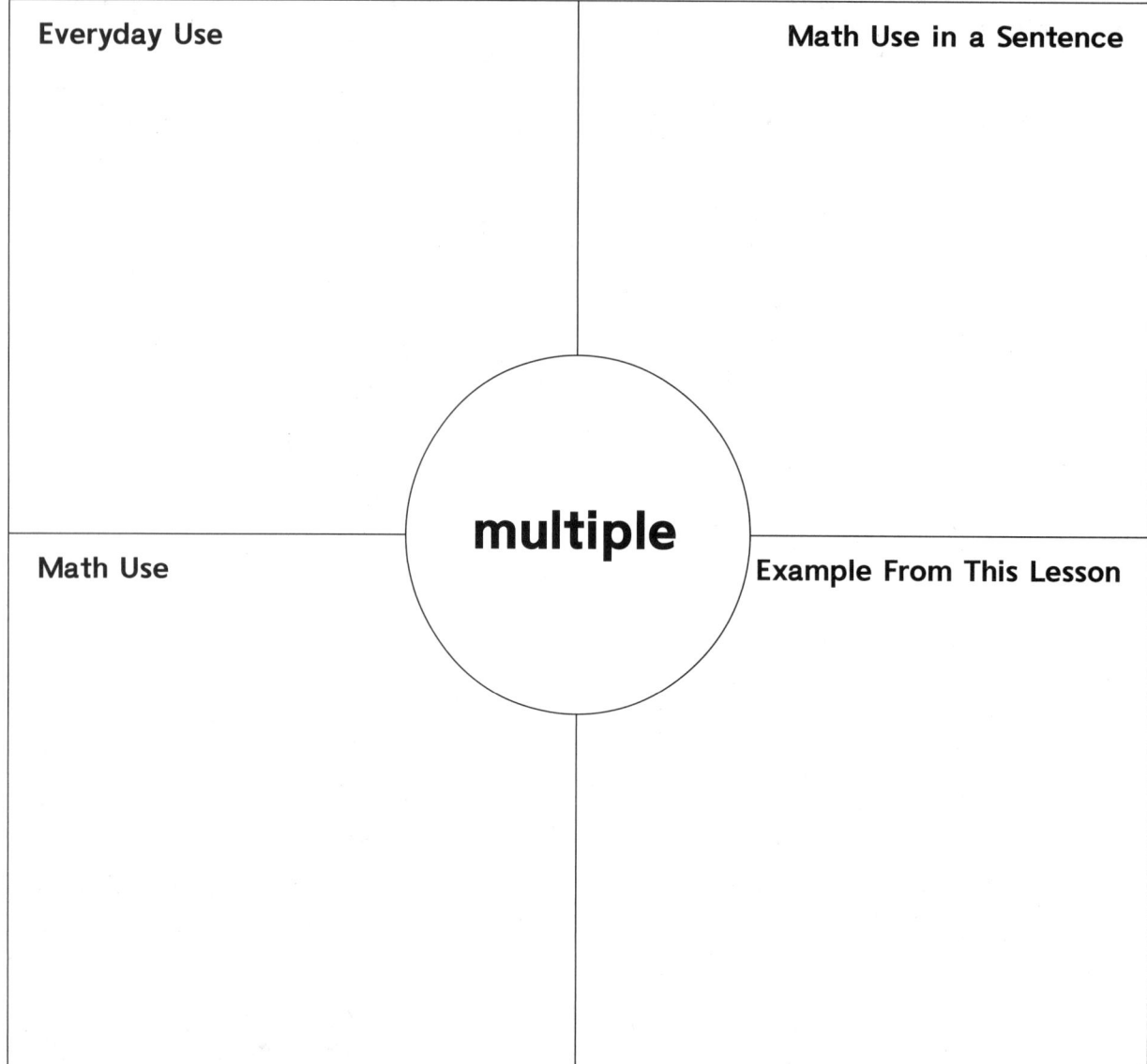

Write the correct numbers on each line to complete the sentences.

Multiples of ____ are 10, 100, 1,000. If 24 ÷ 8 = 3, then 240 ÷ 8 = ____.

**26** Grade 5 • Chapter 3 *Divide by a One-Digit Divisor*

NAME _____ DATE _____

# Lesson 5 Vocabulary Definition Map
*Estimate Quotients*

Use the definition map to write a description and list characteristics about the vocabulary word or phrase. Write or draw math examples. Share your examples with a classmate.

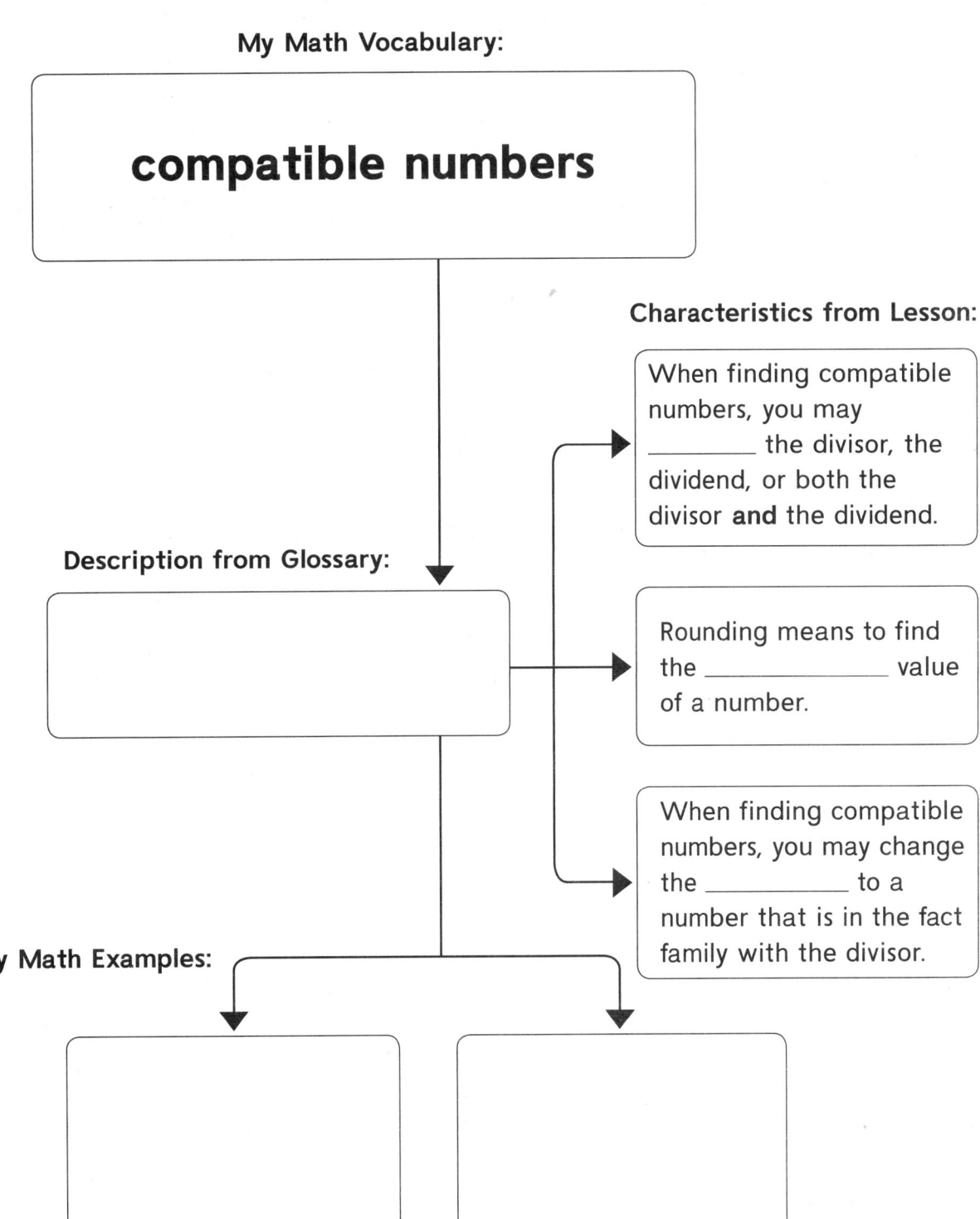

Grade 5 • Chapter 3 *Divide by a One-Digit Divisor* 27

# Lesson 6 Note Taking

## Inquiry/Hands On: Division Models with Greater Numbers

Read the question. Write words you need help with and research each word. Use your lesson to write your Cornell notes. Write or draw math examples to explain your thinking. Share your examples with a classmate.

| **Building on the Essential Question**<br><br>How do you model division with greater numbers? | **Notes:**<br><br>When you model the number 484 using base-ten blocks, you use ____ hundreds flats, ____ tens rods, and ____ single cubes.<br><br>452 ÷ 4 = ?  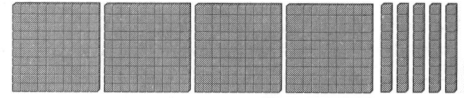<br><br>To model the division, start by dividing the hundreds into ____ equal groups, there will be ____ hundred in each group.<br><br>After dividing the hundreds into 4 equal groups, there will be ____ hundreds remaining.<br><br>If you divide the tens into 4 equal groups, there will be ____ ten in each group.<br><br>After dividing the tens into 4 equal groups, there will be ____ ten remaining. If you regroup the remaining ____ ten into ones, you will have ____ ones altogether.<br><br>If you divide the 12 ones into 4 equal groups, there will be ____ ones in each group.<br><br>Each group has ____ hundred, ____ ten, and ____ ones.<br><br>452 ÷ 4 = ____ |
|---|---|
| **Words I need help with:** | |
| **My Math Examples:** | |

# Lesson 7 Concept Web

## Inquiry/Hands On: Distributive Property and Partial Quotients

Use the concept web to identify the partial quotients and partial sums involved in the division problem.

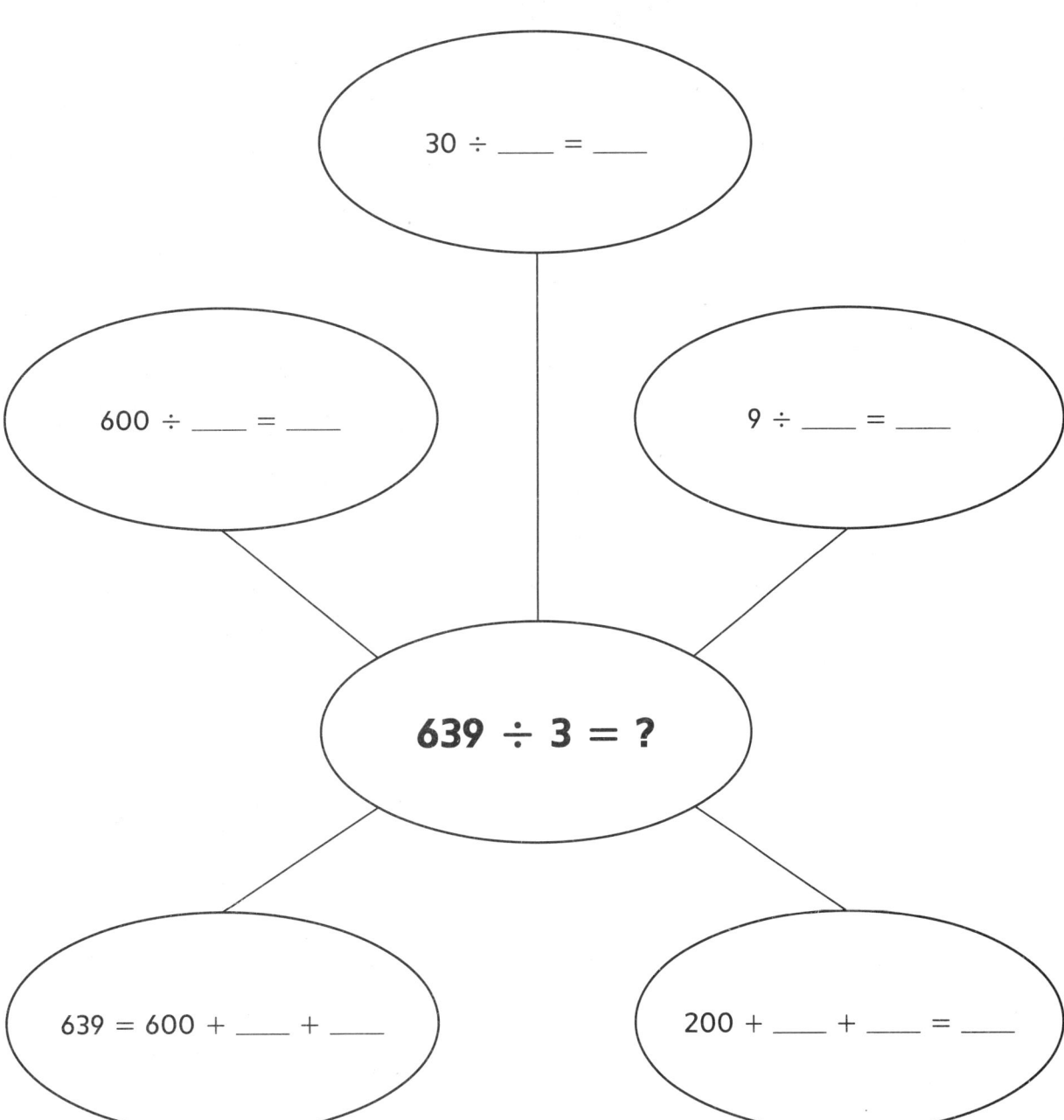

Grade 5 • Chapter 3 *Divide by a One-Digit Divisor*

NAME _____  DATE _____

# Lesson 8 Vocabulary Definition Map

*Divide Three- and Four-Digit Dividends*

Use the definition map to write a description and list characteristics about the vocabulary word or phrase. Write or draw math examples. Share your examples with a classmate.

My Math Vocabulary:

**place value (in division)**

**Characteristics from Lesson:**

When dividing a three- or four-digit number, divide from _____ place value to _____ place value.

Divide each place value by the _____ to find the partial quotients. Multiply the partial quotient and divisor to find the product.

Subtract the _____ from the _____ and compare to the divisor before moving to the next place value.

**Description from Glossary:**

**My Math Examples:**

30 Grade 5 • Chapter 3 *Divide by a One-Digit Divisor*

NAME _____ DATE _____

# Lesson 9 Multiple Meaning Word
*Place the First Digit*

Complete the four-square chart to review the multiple meaning word.

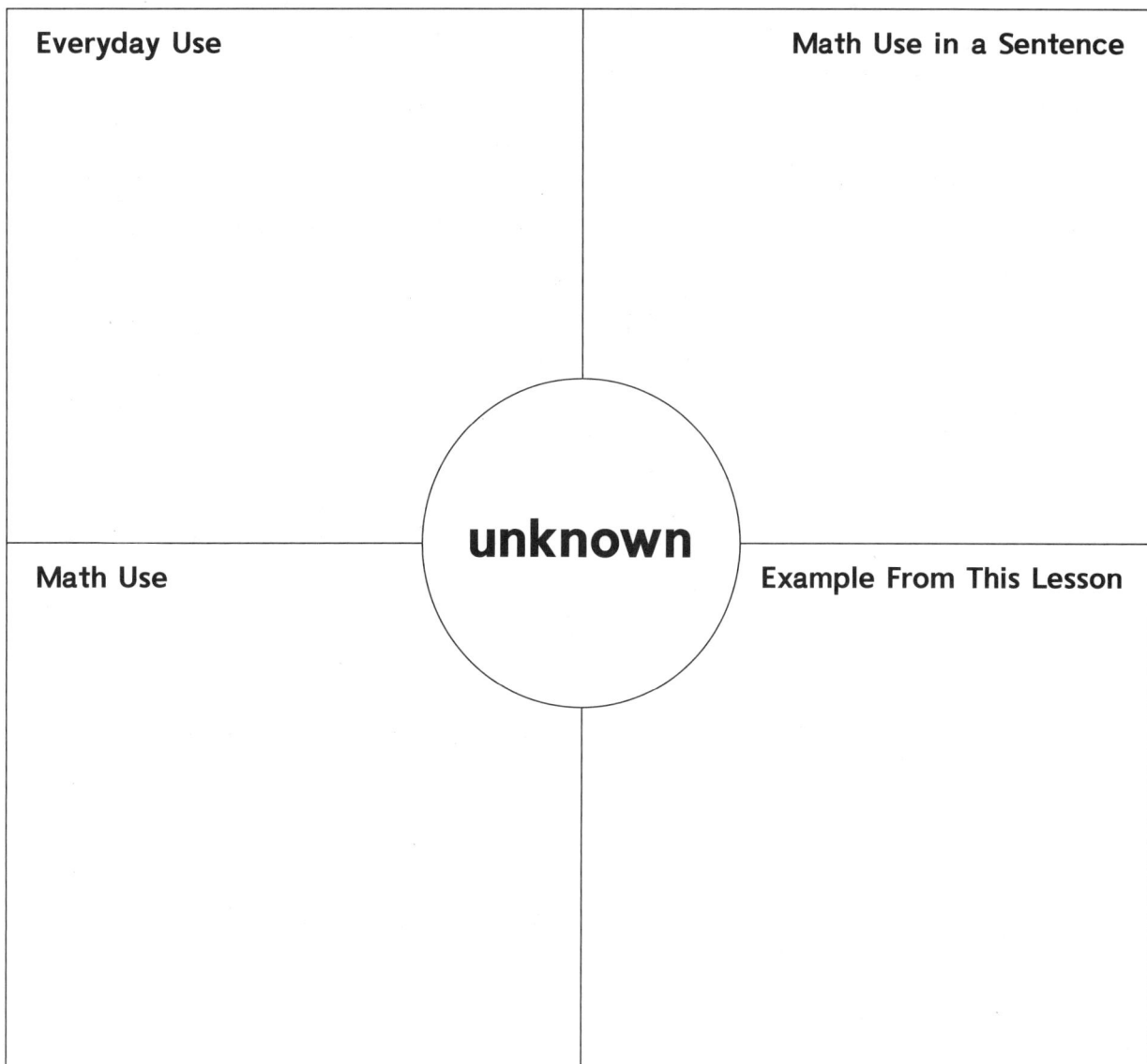

Write the correct term on the line to complete the sentence.

Alexander has 138 cards to organize into sheets that hold 6 cards each. He wants to **know** how many sheets he will need to place every card into a sheet. The unknown is the number of _____.

# Lesson 10 Note Taking
## Quotients with Zeros

Read the question. Write words you need help with and research each word. Use your lesson to write your Cornell notes. Write or draw math examples to explain your thinking. Share your examples with a classmate.

| **Building on the Essential Question** | **Notes:** |
|---|---|
| How do you find quotients with zeros? | $428 \div 4 = ?$ 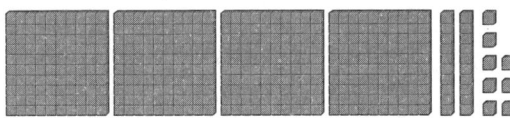 |
| | To model the division, start by dividing the hundreds into ____ equal groups, there will be ____ hundred in each group. |
| | After dividing the hundreds into 4 equal groups, there will be ____ hundreds remaining. |
| | You have ____ tens. If you divide the tens into 4 equal groups, there will be ____ tens in each group. If there are not enough tens to divide, place a ____ in the quotient. |
| **Words I need help with:** | There will be ____ tens remaining. If you regroup the ____ tens into ones, you will have ____ ones altogether. |
| | If you divide the ones into 4 equal groups, there will be ____ ones in each group. |
| | Each group has ____ hundred, ____ tens and ____ ones. |
| | So, $428 \div 4 =$ ____ |

**My Math Examples:**

32  Grade 5 • Chapter 3 *Divide by a One-Digit Divisor*

# Lesson 11 Guided Writing

## Inquiry/Hands On: Use Models to Interpret the Remainder

**How do you model and interpret the remainder?**

Use the exercises below to help you build on answering the Essential Question. Write the correct word or phrase on the lines provided.

1. Rewrite the question in your own words.
   _____
   _____

2. What key words do you see in the question?
   _____

3. There are 75 people on a field trip; a row on the trolley seats 6 people. If each row is filled completely with people from the field trip before moving to the next row, how many rows will have people from the field trip sitting on them? Identify the division expression that will solve the problem. _____

   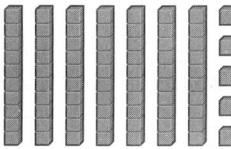

4. If you divide the tens into 6 equal groups, there will be ____ ten in each group.

5. After dividing the tens into 6 equal groups, there will be ____ ten remaining. If you regroup the remaining ____ ten into ones, you would have ____ ones altogether.

6. If you divide the 15 ones into 6 equal groups, there will be ____ ones in each group AND there will be ____ ones remaining.

7. How many rows will be filled completely with people from the field trip? How many will be filled partially?
   _____

8. So, ____ + ____ = ____ describes the number of rows that will have people from the field trip sitting on them.

9. How do you model and interpret the remainder?
   _____
   _____

Grade 5 • Chapter 3 Divide by a One-Digit Divisor

NAME _____ DATE _____

# Lesson 12 Note Taking

*Interpret the Remainder*

Read the question. Write words you need help with and research each word. Use your lesson to write your Cornell notes. Write or draw math examples to explain your thinking. Share your examples with a classmate.

| Building on the Essential Question | Notes: |
|---|---|
| How can you interpret the remainder of a division problem? | The fifth grade classes collected 347 cans for the local food bank. The cans were packed into four boxes in a way that each box contains the same number of cans.<br><br>347 ÷ 4 = ?<br><br>Start by dividing the hundreds into ____ equal groups, there will be ____ hundreds in each group.<br><br>Regroup the ____ hundreds into tens and you have ____ tens altogether.<br><br>If you divide the tens into 4 equal groups, there will be ____ tens in each group, with ____ tens remaining.<br><br>Regroup the remaining ____ tens into ones and you have ____ ones altogether.<br><br>If you divide the ones into 4 equal groups, there will be ____ ones in each group with ____ ones remaining.<br><br>347 ÷ 4 = ____ R ____<br><br>Each box contains ____ cans and there are ____ cans **remaining** that are not in a box. |
| **Words I need help with:** | |

**My Math Examples:**

NAME _____ DATE _____

# Lesson 13 Problem-Solving Investigation

*STRATEGY: Determine Extra or Missing Information*

**Determine if there is extra or missing information. Then solve the problem, if possible.**

1. **Jayden** is downloading songs onto **his** MP3 player. One song is **5 minutes** long, another is **2 minutes** long, and a third is **between** the lengths of the other two songs. What is the **total** length of **all three** songs?

| Understand | Solve |
|---|---|
| I know: | |
| I need to find: | |
| **Plan** | **Check** |
| Facts that are important to solve the problem are: | |
| The extra or missing information is: | |

2. Room 220 and Room 222 are having a canned food drive. **Room 222** collected **346 cans** and **Room 220** collected **278 cans**. How many **more** cans has Room 222 collected **than** Room 220?

| Understand | Solve |
|---|---|
| I know: | |
| I need to find: | |
| **Plan** | **Check** |
| Facts that are important to solve the problem are: | |
| The extra or missing information is: | |

Grade 5 • Chapter 3 Divide by a One-Digit Divisor

NAME _____ DATE _____

# Chapter 4 Divide by a Two-Digit Divisor

*Inquiry of the Essential Question:*

**What strategies can I use to divide by a two-digit number?**

Read the Essential Question. Describe your observations (I see...), inferences (I think...), and prior knowledge (I know...) of each math example. Write additional questions you have below. Then share your ideas and questions with a classmate.

```
      26
14)364        Step 1 Divide the tens.
 -28          Step 2 Multiply, subtract,
  84                 and compare.
 -84          Step 3 Bring down the ones.
   0          Step 4 Divide the ones.
```

I see ...

I think...

I know...

Each group contains 1 ten and 1 one.
So, 132 ÷ 12 = 11.

I see ...

I think...

I know...

**Use rounding to adjust the quotient with a two-digit divisor.**

**Step 1** Use compatible numbers to estimate.
**Step 2** Try the estimate.
**Step 3** Adjust the estimate as needed.
**Step 4** Divide using the division algorithm.

I see ...

I think...

I know...

Questions I have...

_____

_____

_____

36  Grade 5 • Chapter 4 *Divide by a Two-Digit Divisor*

NAME _____ DATE _____

# Lesson 1 Note Taking

## Estimate Quotients

Read the question. Write words you need help with and research each word. Use your lesson to write your Cornell notes. Write or draw math examples to explain your thinking. Share your examples with a classmate.

| **Building on the Essential Question** | **Notes:** |
|---|---|
| How do you estimate quotients? | When you _____ a number, you find the approximate value of a number. |
| | 37 rounded to the nearest ten is _____. |
| | 849 rounded to the nearest hundred is _____. |
| | When you find an _____, you find a number close to an exact value. |
| | Using **rounded numbers,** an estimate for 849 ÷ 37 is _____ ÷ _____ = _____. |
| **Words I need help with:** | _____ numbers are numbers in a problem that are easy to work with mentally. |
| | 8 ÷ 4 is ____ <br> 4 ÷ 4 is ____ <br> 84 ÷ 4 is ____ |
| | Using **compatible numbers,** an estimate for 849 ÷ 37 is _____ ÷ _____ = _____. |
| **My Math Examples:** ||

Grade 5 • Chapter 4 *Divide by a Two-Digit Divisor*

NAME _____   DATE _____

# Lesson 2 Guided Writing

*Inquiry/Hands On: Divide Using Base-Ten Blocks*

**How do you model division using base-ten blocks?**

Use the exercises below to help you build on answering the Essential Question. Write the correct word or phrase on the lines provided.

1. Rewrite the question in your own words.

   _____

   _____

2. What key words do you see in the question?

   _____

3. What numbers are modeled with each base-ten block example below?

      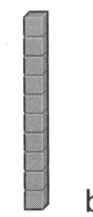

   a. ____          b. ____

4. How would you model 168 using the least amount of base-ten blocks?
   ____ hundred, ____ tens, and ____ ones

5. You cannot divide the hundreds into 12 equal groups. So, regroup the hundreds into tens, and you have ____ tens altogether.

6. Divide the tens into 12 equal groups. There will be ____ ten in each group. Regroup the remaining ____ tens into ones, and you will have ____ ones altogether.

7. Divide the ones into 12 equal groups. There will be ____ ones in each group.

8. 168 ÷ 12 = ____

9. How do you model division?

   _____

   _____

   _____

**38** Grade 5 • Chapter 4 *Divide by a Two-Digit Divisor*

# Lesson 3 Multiple Meaning Word
*Divide by a Two-Digit Divisor*

Complete the four-square chart to review the multiple meaning word.

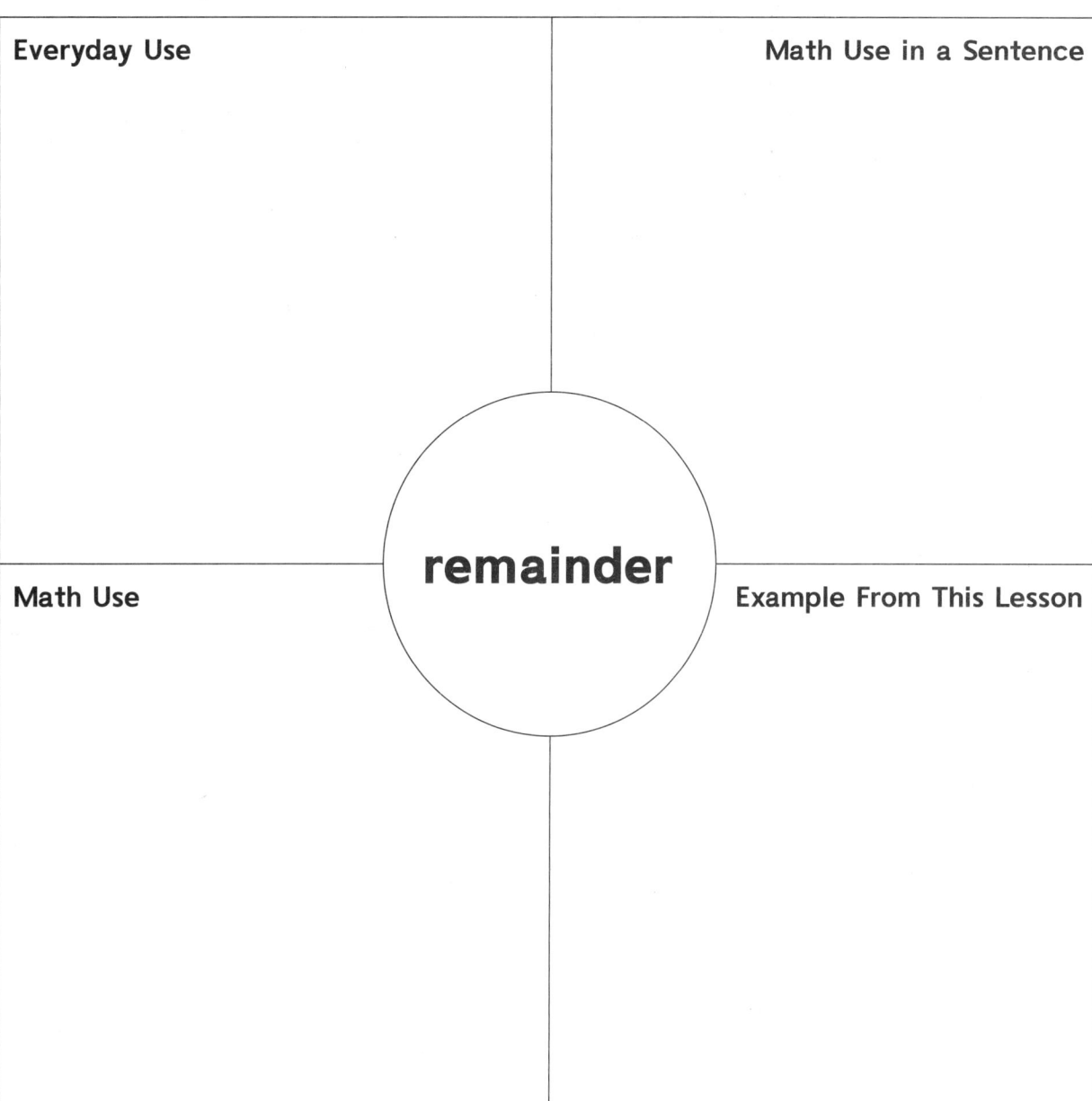

Write the correct term on the line to complete the sentence.

The remainder of a division problem will always be _____ than the divisor.

# Lesson 4 Vocabulary Cognates
## *Adjust Quotients*

Use the Glossary to define the math word in English and in Spanish in the word boxes. Write a sentence using your math word.

| estimate | estimación |
|---|---|
| Definition | Definición |

My math word sentence:

| quotient | cociente |
|---|---|
| Definition | Definición |

My math word sentence:

# Lesson 5 Concept Web

*Divide Greater Numbers*

Use the concept web to identify the parts of a division sentence.

| Word Bank |
|---|
| quotient    dividend    divisor    remainder |

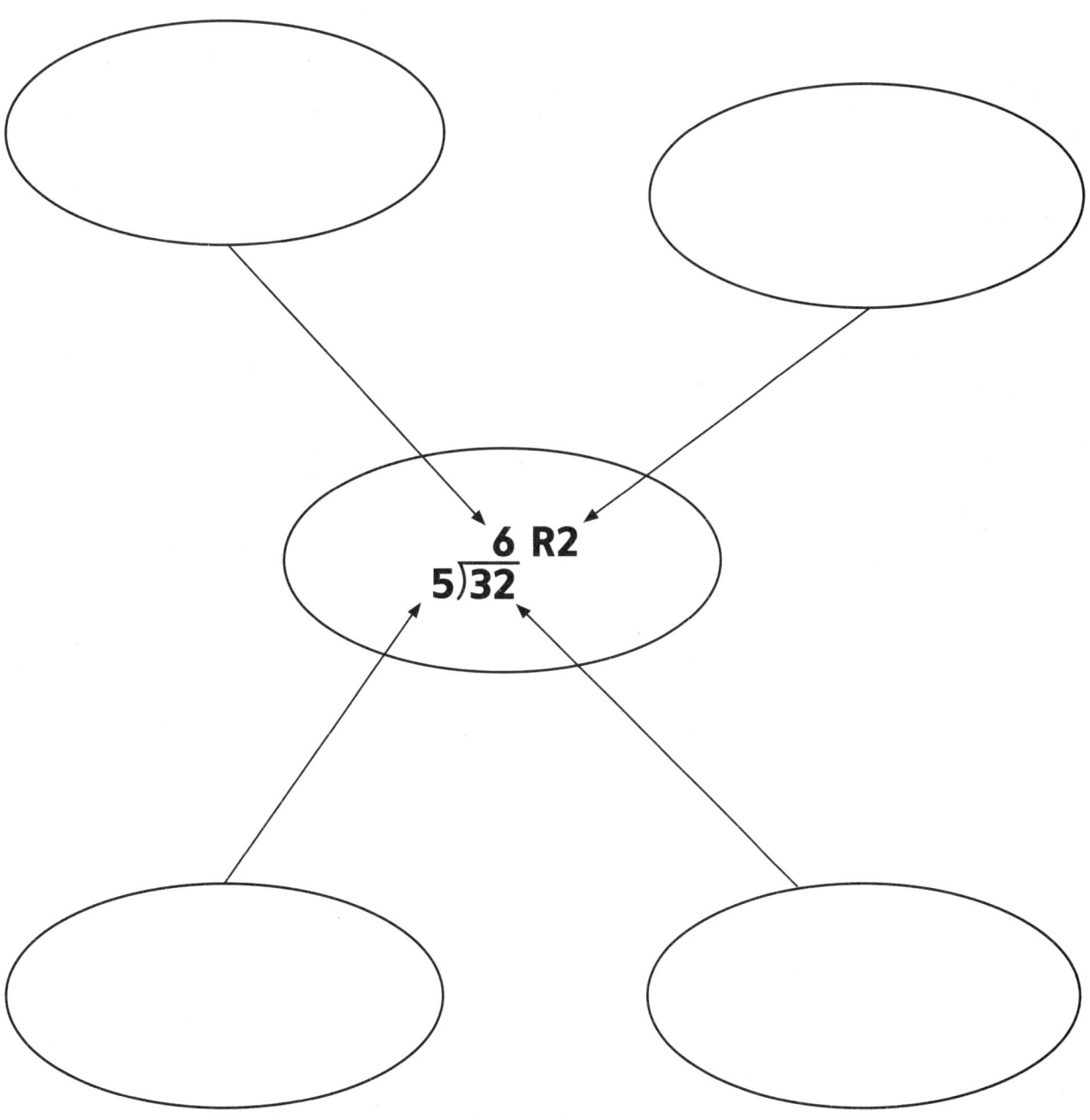

Grade 5 • Chapter 4 *Divide by a Two-Digit Divisor*

# Lesson 6 Problem-Solving Investigation

**STRATEGY:** *Solve a Simpler Problem*

Using a simpler problem, solve each problem below.

1. Mr. Santiago has a flight from New York to Paris that covers a distance of **3,640** **miles** in **7 hours**. If the **plane** travels at the same speed per hour, how many **miles** will **it** have traveled after **4 hours**?

plane

| Understand | Solve |
|---|---|
| I know: | |
| I need to find: | |
| **Plan** | **Check** |
| Speed plane traveled in 1 hour: | |
| Speed plane traveled in 4 hours: | |

2. **Josh** watches **720** television **shows** in **one year**. If **he** watches the **same** number of shows **each month**, how many shows does he watch in **5 months**?

television show

| Understand | Solve |
|---|---|
| I know: | |
| I need to find: | |
| **Plan** | **Check** |
| Shows watched in 1 month: | |
| Shows watched in 5 months: | |

42 Grade 5 • Chapter 4 Divide by a Two-Digit Divisor

# Chapter 5 Add and Subtract Decimals

## Inquiry of the Essential Question:

**How can I use place value and properties to add and subtract decimals?**

Read the Essential Question. Describe your observations (I see...), inferences (I think...), and prior knowledge (I know...) of each math example. Write additional questions you have below. Then share your ideas and questions with a classmate.

| 8.72 — rounds to → 8.7 | I see ... |
|---|---|
| − 3.05 — rounds to → − 3.1 | |
|                   5.6 | I think... |
| | I know... |

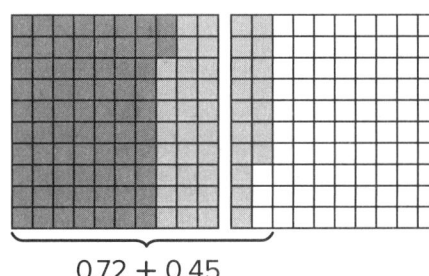

0.72 + 0.45

0.72 + 0.45 = 1.17

I see ...

I think...

I know...

| 1.5 + 6.4 + 3.5 = 1.5 + 3.5 + 6.4 | Commutative Property | I see ... |
|---|---|---|
| = (1.5 + 3.5) + 6.4 | Associative Property | I think... |
| = 5 + 6.4 | Add. | |
| = 11.4 | Add. | I know... |

Questions I have...

_____

_____

_____

Grade 5 • Chapter 5 *Add and Subtract Decimals*

NAME _____ DATE _____

# Lesson 1 Vocabulary Cognates

*Round Decimals*

Use the Glossary to define the math word in English and in Spanish in the word boxes. Write a sentence using your math word.

| decimal | decimal |
|---|---|
| Definition | Definición |
| My math word sentence: | |

| rounding | redondear |
|---|---|
| Definition | Definición |
| My math word sentence: | |

44 Grade 5 • Chapter 5 *Add and Subtract Decimals*

# Lesson 2 Vocabulary Definition Map
## *Estimate Sums and Differences*

Use the definition map to write a description and list characteristics about the vocabulary word or phrase. Write or draw math examples. Share your examples with a classmate.

My Math Vocabulary:

**estimate**

Description from Glossary:

Characteristics from Lesson:

When estimating with decimals, you could _____ the decimal to the nearest ten or one.

When rounding, if the digit to the right of the place value you are rounding is 4 or **less**, round _____.

When rounding, if the digit to the right of the place value you are rounding is 5 or **greater**, round _____.

My Math Examples:

Grade 5 • Chapter 5 *Add and Subtract Decimals* **45**

NAME _____ DATE _____

# Lesson 3 Problem-Solving Investigation

## STRATEGY: *Estimate or Exact Answer*

Determine whether you need an estimate or exact answer to solve each problem.

1. A restaurant can make **95** dinners <u>each</u> night. The restaurant has been **sold out** for <u>7 nights</u> in a row. How many dinners were sold during <u>this</u> week?

| Understand | Solve |
|---|---|
| I know: | |
| I need to find: | |
| **Plan** | **Check** |
| There are 7 nights in 1 week. **Sold out** means, "all 95 dinners were sold that night." | |

2. A **family** is renting a cabin for **$59.95** <u>a</u> <u>day</u> for **3** days. <u>About</u> how much will **they** (the family) pay for the cabin?

| Understand | Solve |
|---|---|
| I know: | |
| I need to find: | |
| **Plan** | **Check** |
| How much will 1 day cost? <u>About</u> how much will 1 day cost? <u>About</u> how much will 3 days cost? | |

46 Grade 5 • Chapter 5 *Add and Subtract Decimals*

# Lesson 4 Guided Writing

## Inquiry/Hands On: Add Decimals Using Base-Ten Blocks

**How do you add decimals using base-ten blocks?**

Use the exercises below to help you build on answering the Essential Question. Write the correct word or phrase on the lines provided.

1. Rewrite the question in your own words.
   _____
   _____

2. What key words do you see in the question?
   _____

3. When using base-ten blocks, which decimals are modeled below, 1.0, 0.1, or 0.01?

   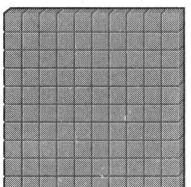 a. _____     b. _____     c. _____

4. When adding two decimals, you can model each decimal using base-ten blocks, and then _____ the base-ten blocks.

5. When you combine two decimals, both modeled using base-ten blocks, you may need to regroup. _____ hundredths blocks (flats) can be regrouped as 1 _____ block (rod). 10 _____ blocks (rods) can be regrouped as 1 _____ block (ones cube).

6. How do you add using base-ten blocks?
   _____
   _____

NAME _____ DATE _____

# Lesson 5 Note Taking

## *Inquiry/Hands On: Add Decimals Using Models*

Read the question. Write words you need help with and research each word. Use your lesson to write your Cornell notes. Write or draw math examples to explain your thinking. Share your examples with a classmate.

| Building on the Essential Question | Notes: |
|---|---|
| How do you model adding decimals? | You can model decimals by _____ squares in a 10-by-10 grid. |

The decimal _____ is modeled by shading one square.

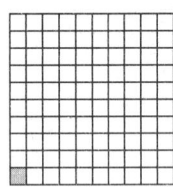

The decimal _____ is modeled by shading ten squares.

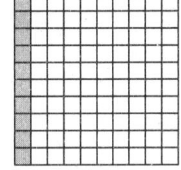

**Words I need help with:**

The decimal _____ is modeled by shading all the squares.

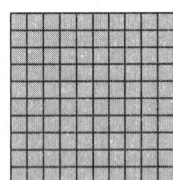

When adding two decimals, you can model each decimal by _____ squares on 10-by-10 grids, and then count the _____ squares to find the _____.

**My Math Examples:**

48  Grade 5 • Chapter 5 *Add and Subtract Decimals*

# Lesson 6 Concept Web

## Add Decimals

Use the concept web to identify the place values of the decimal.

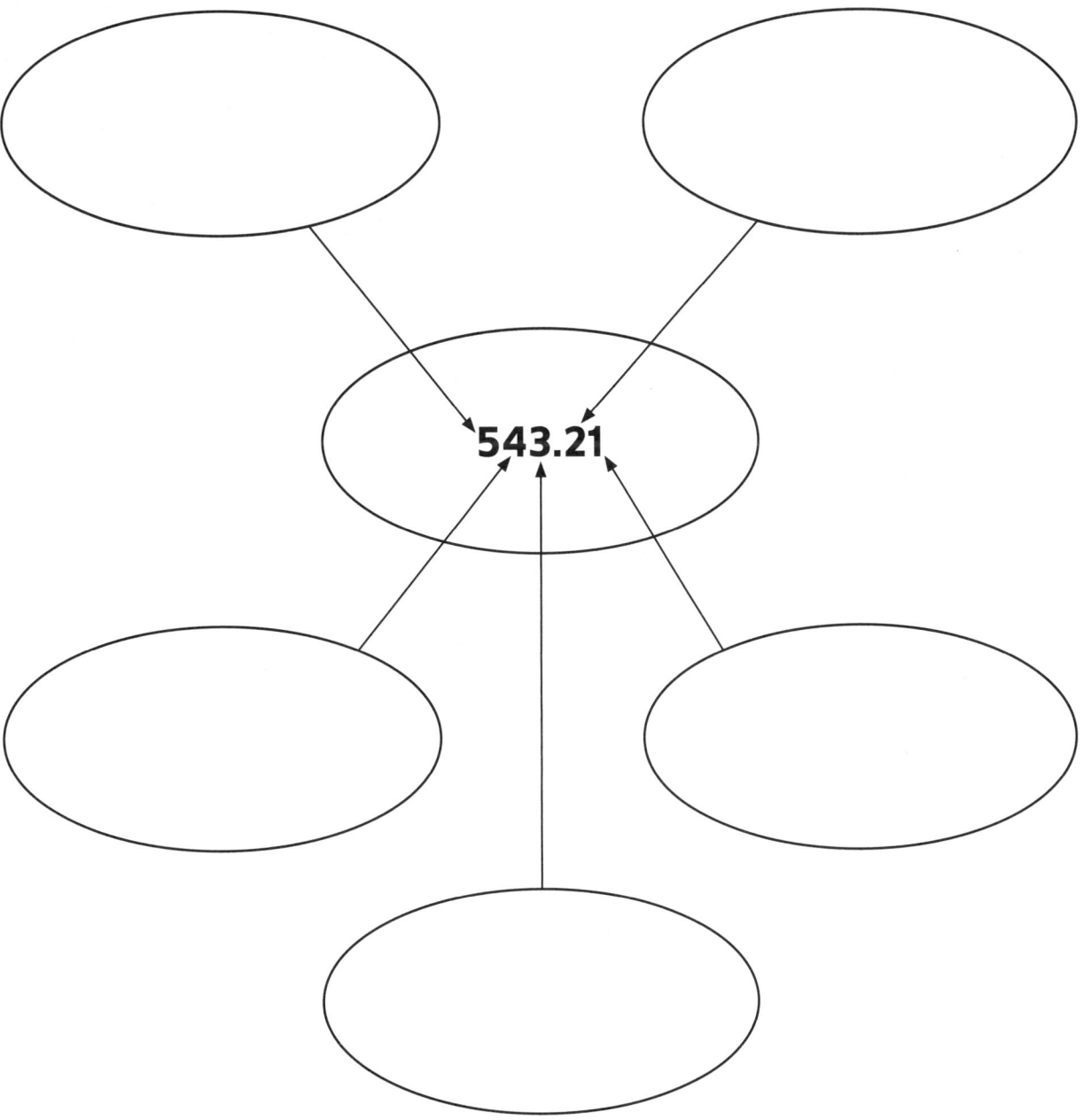

Grade 5 • Chapter 5 *Add and Subtract Decimals*

NAME _____  DATE _____

# Lesson 7 Vocabulary Chart
*Addition Properties*

Use the three-column chart to organize the vocabulary in this lesson. Write the word in Spanish. Then write the correct terms to complete each definition.

| English | Spanish | Definition |
|---|---|---|
| **property** | | A _____ in mathematics that can be applied to all numbers. |
| **Associative Property** | | Property that states that the way in which numbers are _____ does not change the sum. |
| **Commutative Property** | | Property that states that the _____ in which numbers are added does not change the _____. |
| **Identity Property** | | Property that states that the sum of any number and 0 _____ the number. |

50  Grade 5 • Chapter 5 *Add and Subtract Decimals*

# Lesson 8 Note Taking

## Inquiry/Hands On: Subtract Decimals Using Base-Ten Blocks

Read the question. Write words you need help with and research each word. Use your lesson to write your Cornell notes. Write or draw math examples to explain your thinking. Share your examples with a classmate.

**Building on the Essential Question**

How do you model subtraction using base-ten blocks?

**Words I need help with:**

**Notes:**

The decimal _____ is modeled below.

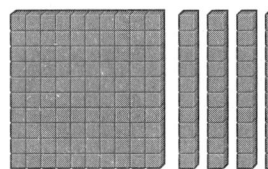

The decimal 0.45 is modeled using ___ tenths blocks (rods) and ___ hundredths blocks (ones cubes).

When you subtract two decimals, start by removing base-ten blocks from the _____ place value.

1.63 − 0.45

There are not 5 hundredths in the base-ten blocks for 1.63. So, you regroup 1 _____ block (rod) as ___ hundredths blocks (ones cubes).

After regrouping, there will be ___ hundredths blocks (ones cubes) altogether.

Subtract ___ hundredths, and there are ___ hundredths remaining.

Next, subtract the tenths place value. After the regrouping, there are ___ tenths blocks (rods).

Subtract ___ tenths, and ___ tenth remains.

After subtracting 0.45 from 1.63, ___ one, ___ tenth, and ___ hundredths remain.

**My Math Examples:**

# Lesson 9 Guided Writing

## Inquiry/Hands On: Subtract Decimals Using Models

**How do you model decimal subtraction?**

Use the exercises below to help you build on answering the Essential Question. Write the correct word or phrase on the lines provided.

1. What key words do you see in the question?
   _____

2. The decimal _____ is modeled in the shaded squares below.

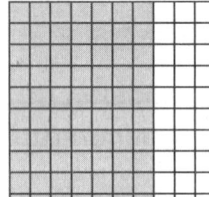

3. When subtracting a decimal from 1.7, you _____ _____ enough squares that represent the decimal on the model for 1.7.

4. To subtract 0.45, cross out _____ squares.

5. The _____ is represented by the number of shaded squares that remain.

6. 1.7 − 0.45 = _____

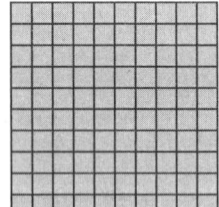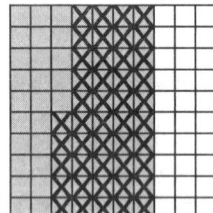

7. How do you model decimal subtraction?
   _____
   _____
   _____
   _____

52 Grade 5 • Chapter 5 *Add and Subtract Decimals*

# Lesson 10 Concept Web

## Subtract Decimals

Use the concept web to identify the inverse operation of each operation shown in the concept web.

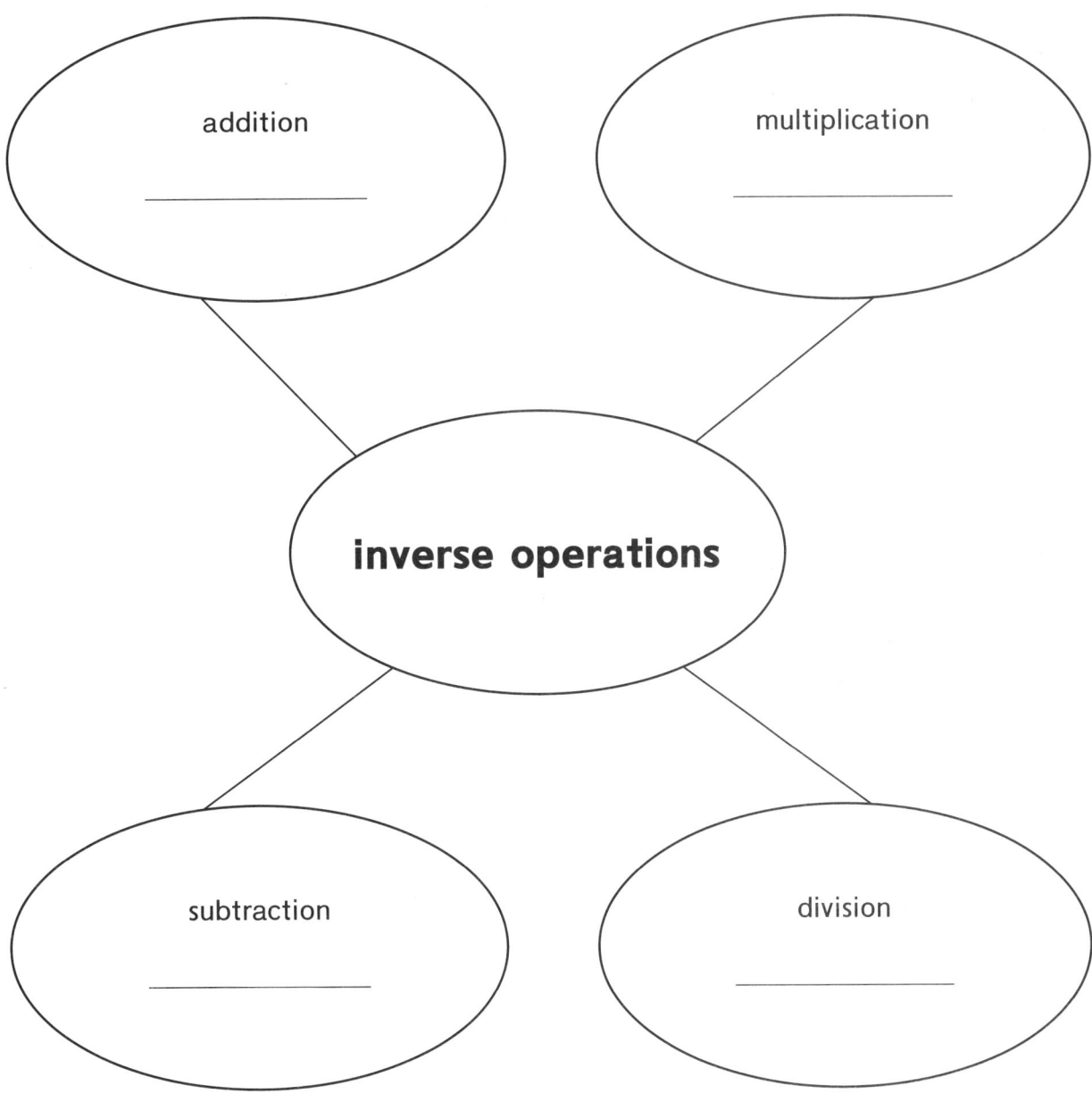

Grade 5 • Chapter 5 *Add and Subtract Decimals* **53**

NAME _____ DATE _____

# Chapter 6 Multiply and Divide Decimals

*Inquiry of the Essential Question:*

**How is multiplying and dividing decimals similar to multiplying and dividing whole numbers?**

Read the Essential Question. Describe your observations (I see..), inferences (I think...), and prior knowledge (I know...) of each math example. Write additional questions you have below. Then share your ideas and questions with a classmate.

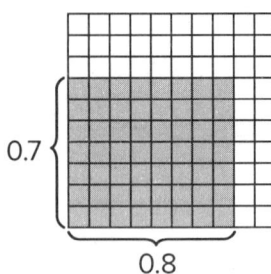

I see ...

I think...

I know...

```
      1
    0.1 3   ← 2 decimal places
 ×   2.5    ← 1 decimal place
   ─────
     6 5
 + 2 6 0
   ─────
   0.3 2 5  ← 2 + 1 = 3 decimal places
```

I see ...

I think...

I know...

34.7  ÷  6
 ↓       ↓
 36   ÷  6   ← Round 34.7 to 36 since 36 and 6 are compatible numbers.

Since 36 ÷ 6 = 6, 34.7 ÷ 6 is **about** 6

I see ...

I think...

I know...

Questions I have...

_____

_____

_____

54 Grade 5 • Chapter 6 *Multiply and Divide Decimals*

NAME _____ DATE _____

# Lesson 1 Review Vocabulary Chart
*Estimate Products of Whole Numbers and Decimals*

Use the three-column chart to organize the review vocabulary in this lesson. Write the word in Spanish. Then write the correct terms to complete each definition.

| English | Spanish | Definition |
|---------|---------|------------|
| **decimal** | | A number that has a _____ in the tenths place, hundredths place, and beyond. |
| **estimate** | | A number _____ to an exact value. An estimate indicates _____ how much. |
| **place value** | | The value given to a digit by its _____ in a number. |
| **product** | | The answer to a _____ problem. |

Grade 5 • Chapter 6 *Multiply and Divide Decimals*  55

NAME _____ DATE _____

# Lesson 2 Guided Writing

*Inquiry/Hands On: Use Models to Multiply*

**How do you use models to multiply decimals?**

Use the exercises below to help you build on answering the Essential Question. Write the correct word or phrase on the lines provided.

1. Rewrite the question in your own words.
   _____
   _____

2. What key words do you see in the question?
   _____

3. You can model decimals by _____ squares in a 10-by-10 grid. The decimal ____ is modeled in the shaded squares to the right.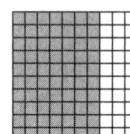

4. To model the multiplication of a whole number and a decimal, use as many grids as the value of the _____ number. Then model the _____ on each grid.

5. To model the multiplication of 0.7 and 3, you will need ____ grids. Model ____ on each of the grids.

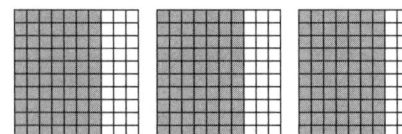

6. To find the product of a whole number and a decimal, combine the shaded squares onto one model. The total number of shaded squares is ____. The model below represents the product, which is ____.

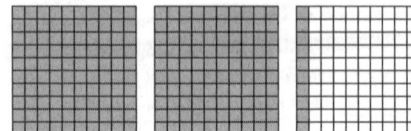

7. How do you model multiplication of whole numbers and decimals?
   _____
   _____
   _____

**56** Grade 5 • Chapter 6 *Multiply and Divide Decimals*

NAME _____ DATE _____

# Lesson 3 Four-Square Vocabulary
## *Multiply Decimals by Whole Numbers*

Write the definition for each review math word. Write what each word means in your own words. Draw or write examples that show each math word meaning. Then write your own sentences using the words.

| Definition | My Own Words |
|---|---|
| | |
| **multiplication** | |
| My Examples | My Sentence |
| | |

| Definition | My Own Words |
|---|---|
| | |
| **factor** | |
| My Examples | My Sentence |
| | |

Grade 5 • Chapter 6 *Multiply and Divide Decimals* **57**

# Lesson 4 Note Taking

*Inquiry/Hands On: Use Models to Multiply Decimals*

Read the question. Write words you need help with and research each word. Use your lesson to write your Cornell notes. Write or draw math examples to explain your thinking. Share your examples with a classmate.

**Building on the Essential Question**

How do you use models to multiply decimals?

**Words I need help with:**

**Notes:**

To model the multiplication of 0.4 and 0.7, shade a _____ that has a _____ of 0.7 and a _____ of 0.4.

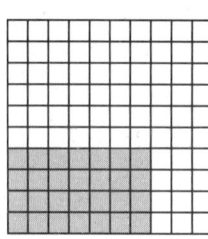

The _____ is found by counting the total number of shaded squares on the rectangle.

The total number of shaded squares for 0.4 × 0.7 is _____ squares. The product is _____.

**My Math Examples:**

# Lesson 5 Vocabulary Note Taking
## *Multiply Decimals*

Read the question. Write words you need help with and research each word. Use your lesson to write your Cornell notes. Write or draw math examples to explain your thinking. Share your examples with a classmate.

| **Building on the Essential Question**<br><br>How do you multiply decimals? | **Notes:**<br><br>Multiplication with decimal numbers is similar to multiplication with _____ numbers.<br><br>When you multiply decimals, _____ as with whole numbers.<br><br>$\phantom{xx}0.25 \phantom{xxxxxx} 25$<br>$\underline{\times \phantom{x} 3.1} \phantom{xxxx} \underline{\times \phantom{x} 31}$<br>$\phantom{xxxxxxxxxxx} 25$<br>$\phantom{xxxxxx} \underline{+ \phantom{x} 750}$<br>$\phantom{xxxxxxxxx} 775$ |
|---|---|
| **Words I need help with:** | Count the decimal places in each decimal factor.<br>There are ___ decimal places in 0.25.<br>There is ___ decimal place in 3.1.<br><br>The total number of decimal places in the factors is _____ to the number of decimal places in the product of the decimals.<br><br>There are a total of ___ decimal places in 0.25 × 3.1.<br>There will be ___ decimal places in the product of 0.25 × 3.1.<br>The decimal point will be placed ___ decimal places from the _____ in 775.<br><br>0.25 × 3.1 = _____ |
| **My Math Examples:** | |

Grade 5 • Chapter 6 *Multiply and Divide Decimals*

# Lesson 6 Concept Web

*Multiply Decimals by Powers of Ten*

Use the concept web to write each exponent as a power of 10 and each power of ten as an exponent.

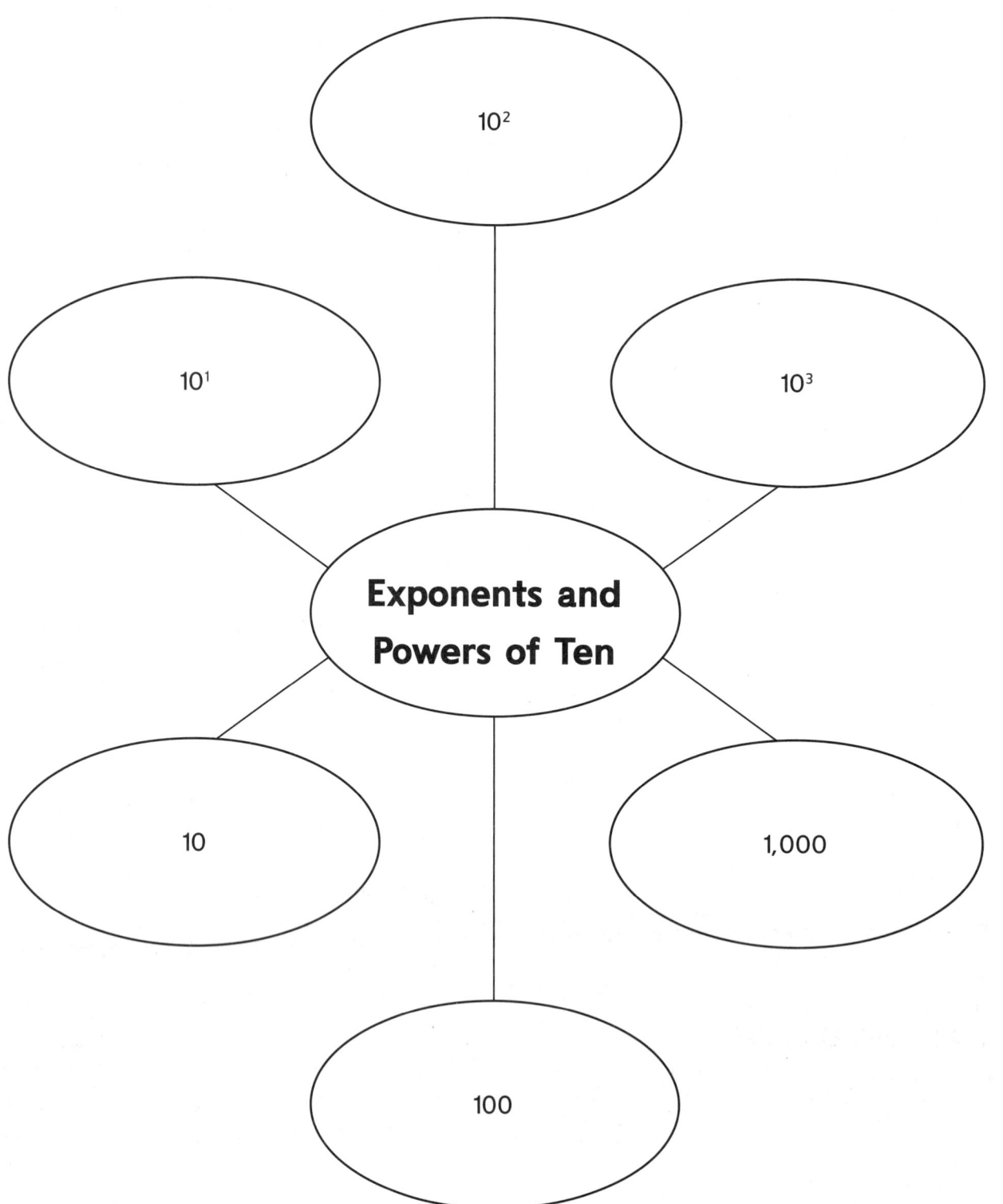

NAME _____ DATE _____

# Lesson 7 Problem-Solving Investigation

## STRATEGY: Look for a Pattern

**Look for a pattern to solve each problem.**

1. Every year, **Victoria** receives **$30** for her birthday, plus **$2** for **each year** of **her** (Victoria's) **age**.
   **Lacey** received **$20** for her birthday and **$4** for **each year** of **her** (Lacey's) **age**.
   In **2013**, Victoria is **10**, and Lacey is **6**.
   In what year will **they** both receive the **same** amount of money?

| Understand | Solve |
| --- | --- |
| I know: <br> I need to find: | <table><tr><th>Year</th><th>Victoria</th><th>Lacey</th></tr><tr><td>2013</td><td>30 + 2 × 10</td><td>20 + 4 × 6</td></tr><tr><td>2014</td><td>30 + 2 × 11</td><td>20 + 4 × 7</td></tr><tr><td>2015</td><td></td><td></td></tr><tr><td>2016</td><td></td><td></td></tr><tr><td>2017</td><td></td><td></td></tr></table> |
| **Plan** <br> I will use a table and look for a _____. | **Check** |

2. **Trent** lifts weights **7 days a week**.
   He spends **18 minutes** lifting weights on **Monday**, 29 minutes on **Tuesday**, **40 minutes** on **Wednesday**, and 51 minutes on **Thursday**.
   If this pattern continues, how **many** minutes will Trent lift weights on **Saturday**?

| Understand | Solve |
| --- | --- |
| I know: <br> I need to find: | |
| **Plan** <br> I will make a _____ and look for a pattern. | **Check** |

Grade 5 • Chapter 6 *Multiply and Divide Decimals* **61**

# Lesson 8 Vocabulary Definition Map
*Multiplication Properties*

Use the definition map to write a description and list characteristics about the vocabulary word or phrase. Write or draw math examples. Share your examples with a classmate.

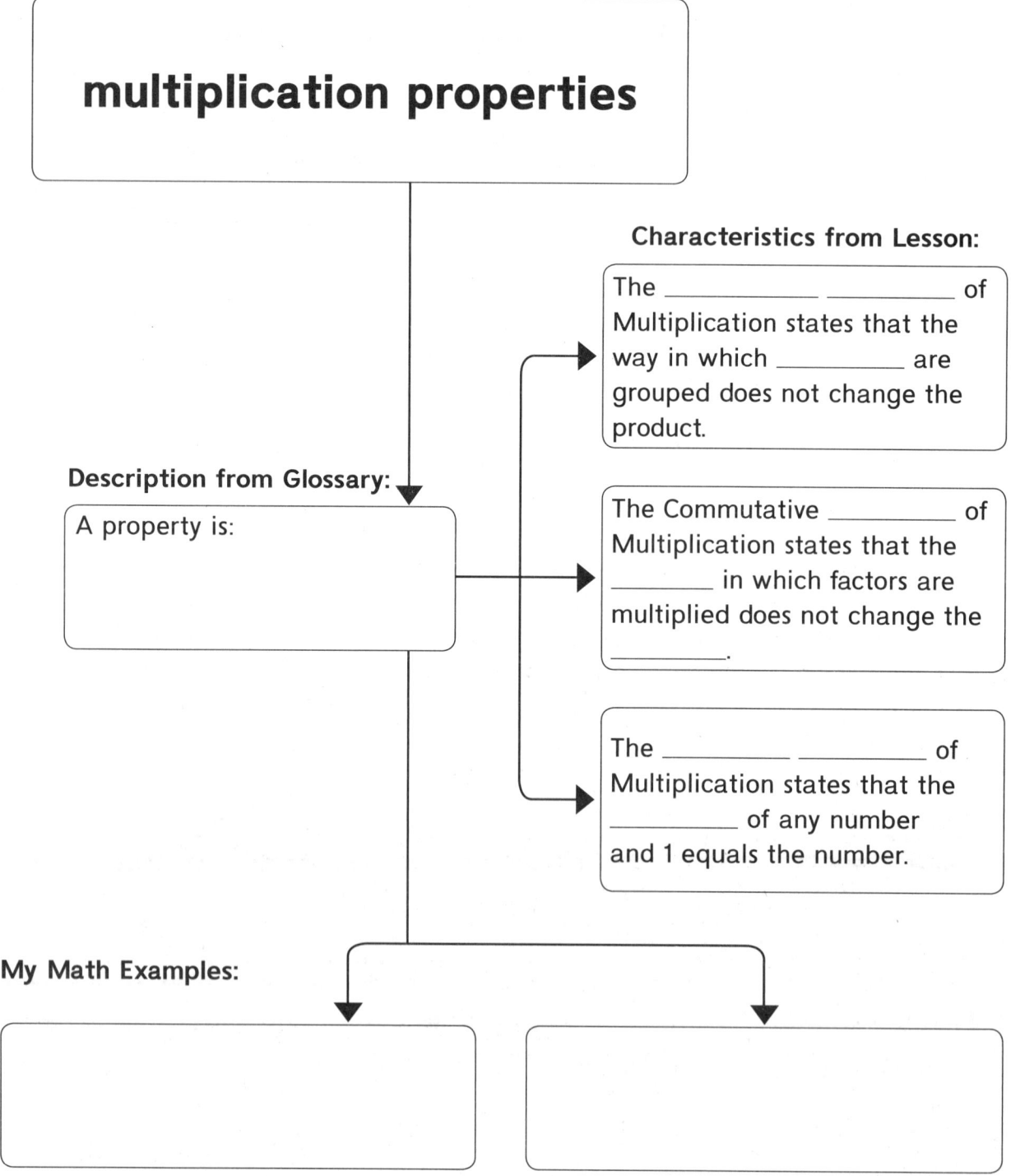

My Math Vocabulary:

**multiplication properties**

**Description from Glossary:**
A property is:

**Characteristics from Lesson:**

The _____ _____ of Multiplication states that the way in which _____ are grouped does not change the product.

The Commutative _____ of Multiplication states that the _____ in which factors are multiplied does not change the _____.

The _____ _____ of Multiplication states that the _____ of any number and 1 equals the number.

**My Math Examples:**

# Lesson 9 Vocabulary Concept Web

*Estimate Quotients*

Use the concept web to identify examples of compatible numbers for each division expression.

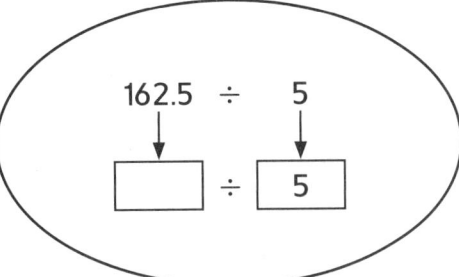

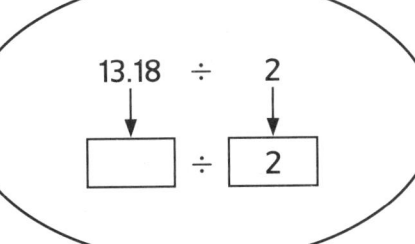

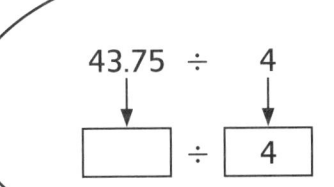

**compatible numbers**

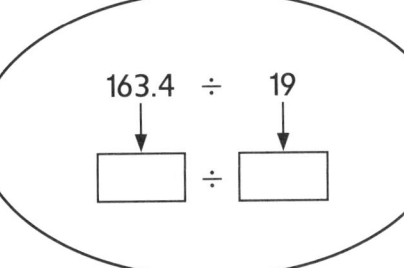

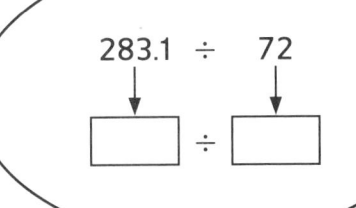

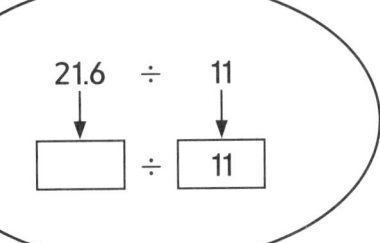

NAME _____ DATE _____

# Lesson 10 Note Taking

*Inquiry/Hands On: Divide Decimals*

Read the question. Write words you need help with and research each word. Use your lesson to write your Cornell notes. Write or draw math examples to explain your thinking. Share your examples with a classmate.

**Building on the Essential Question**

How do you divide decimals using base-ten blocks?

**Words I need help with:**

**Notes:**

The decimal ____ is modeled below using base-ten blocks.

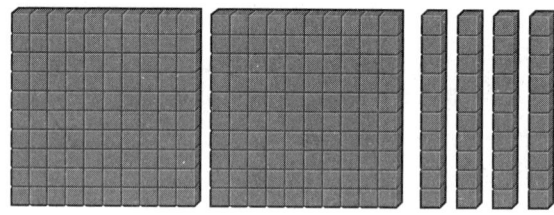

If you divide the whole blocks into 2 equal groups, there will be ____ whole in each group.

If you divide the tenths into 2 equal groups, there will be ____ tenths in each group.

Each group has ____ whole and ____ tenths.

Each group represents the decimal ____.

You have just modeled the following division equation:

____ ÷ 2 = ____

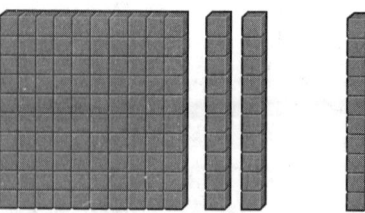    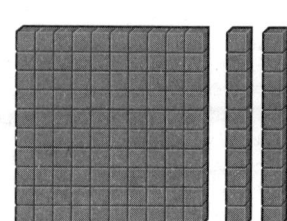

**My Math Examples:**

**64** Grade 5 • Chapter 6 *Multiply and Divide Decimals*

# Lesson 11 Multiple Meaning Word
*Divide Decimals by Whole Numbers*

Complete the four-square chart to review the multiple meaning word or phrase.

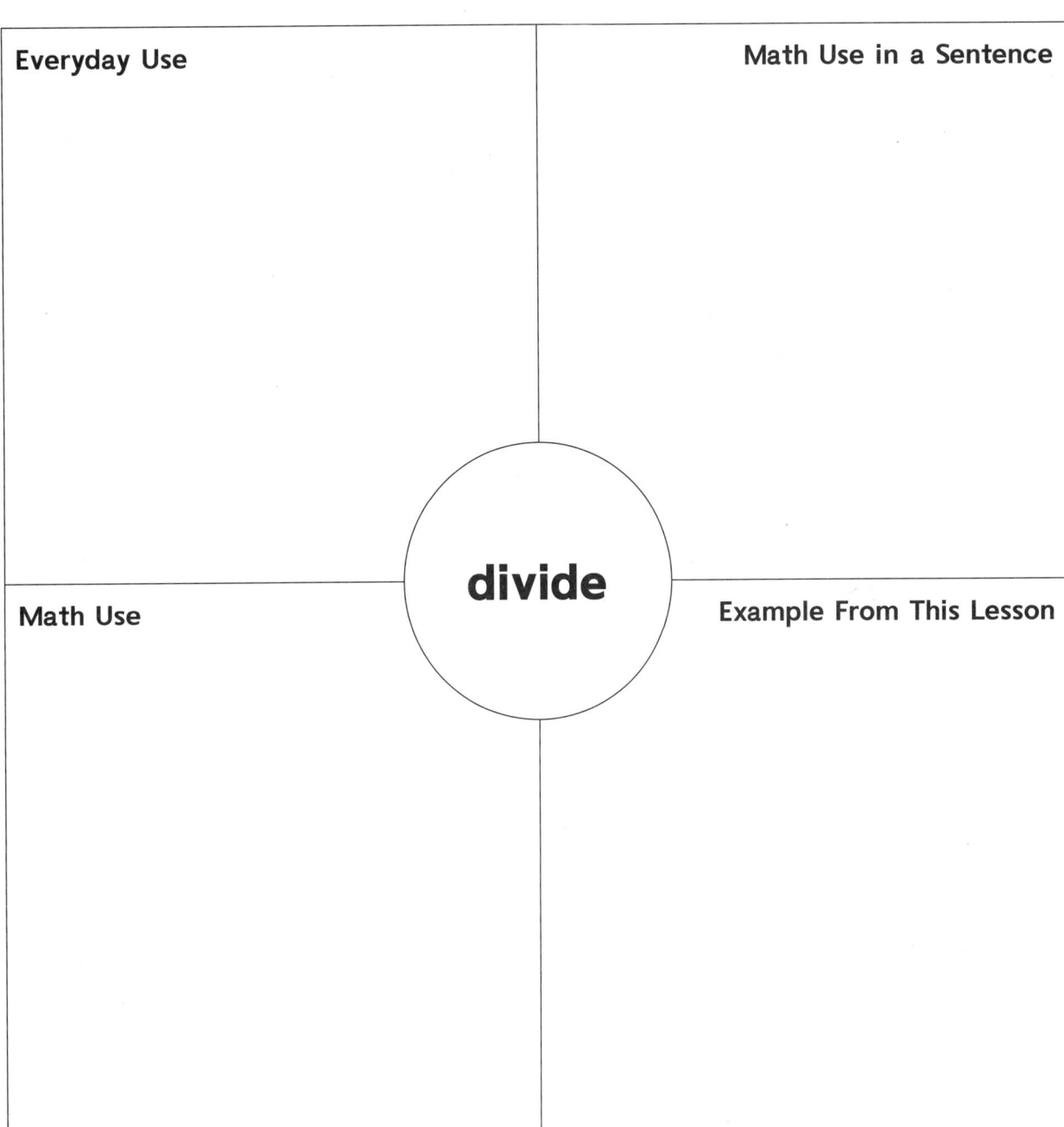

Write the correct term on the line to complete the sentence.

Division of decimals is similar to division of whole numbers except there is a _____ _____ appearing in the dividend and the quotient.

NAME _____ DATE _____

# Lesson 12 Guided Writing

## Inquiry/Hands On: Use Models to Divide Decimals

**How do you model dividing decimals using base-ten blocks?**

Use the exercises below to help you build on answering the Essential Question. Write the correct word or phrase on the lines provided.

1. Rewrite the question in your own words.
   _____
   _____

2. What key words do you see in the question?
   _____

3. What decimal is modeled using the base-ten blocks below?
   _____

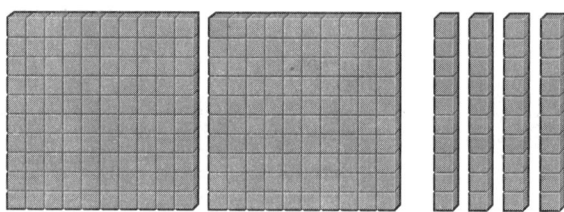

4. If you divide a decimal by tenths, you will regroup the decimal into _____.

5. The decimal 0.6 written in word form is ____ _____. When you divide 2.4 by 0.6, you are dividing by _____.

6. So, you will regroup the _____ in 2.4 into tenths. There are ____ tenths in 2 wholes. After regrouping, there are ____ tenths altogether.

7. If you separate the 24 tenths into groups of 6 tenths, you will have ____ groups.

8. 24 ÷ 6 = ____ and 24 tenths ÷ 6 tenths = ____ tenths. So, 2.4 ÷ 6 = ____

9. How do you model dividing decimals using base-ten blocks?
   _____
   _____
   _____
   _____

**66** Grade 5 • Chapter 6 *Multiply and Divide Decimals*

NAME _____   DATE _____

# Lesson 13 Concept Web

## *Divide Decimals*

Use the concept web to identify the parts of a decimal division sentence.

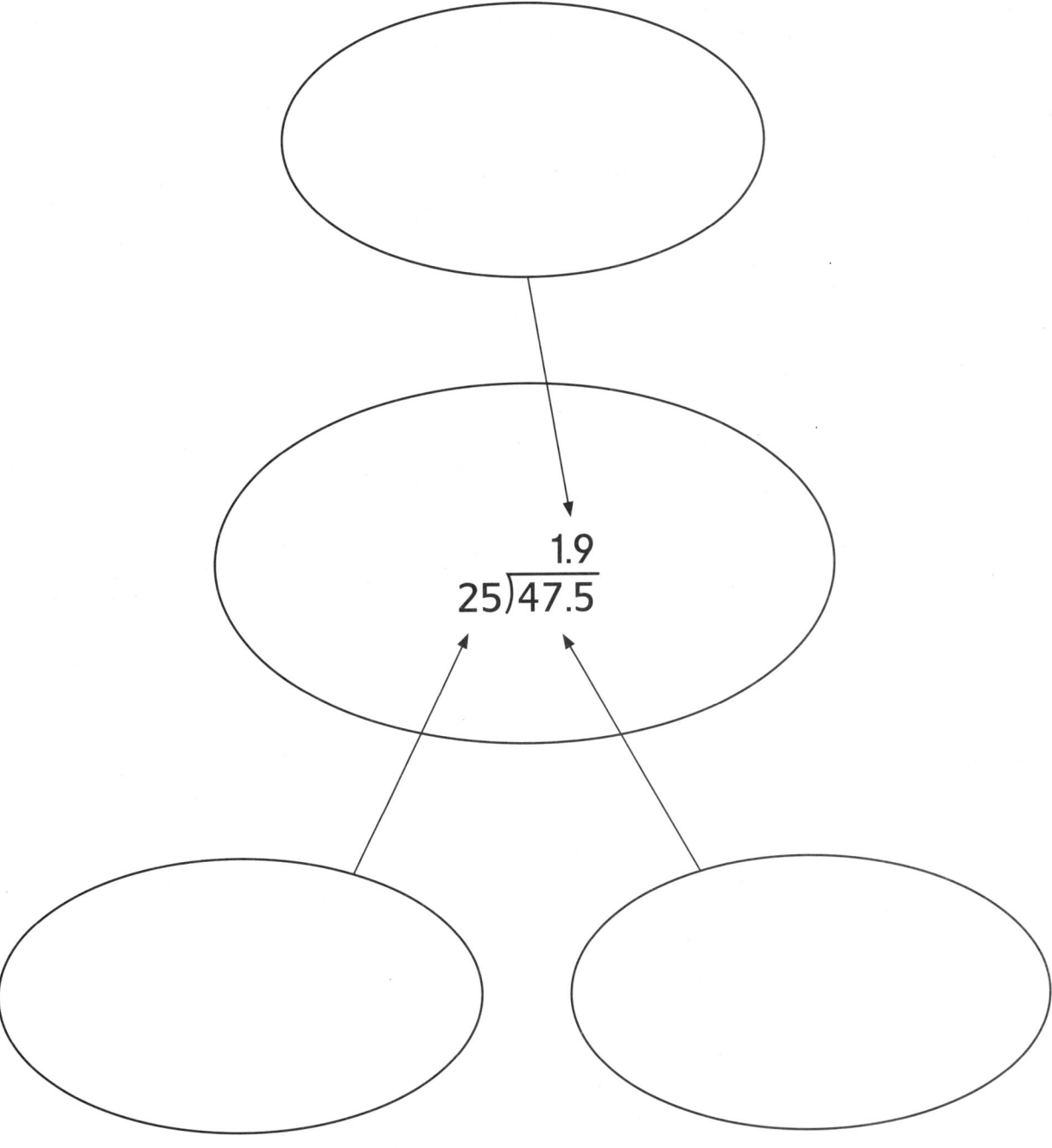

Grade 5 • Chapter 6 *Multiply and Divide Decimals* **67**

# Lesson 14 Vocabulary Definition Map

*Divide Decimals by Powers of Ten*

Use the definition map to write a description and list characteristics about the vocabulary word or phrase. Write or draw math examples. Share your examples with a classmate.

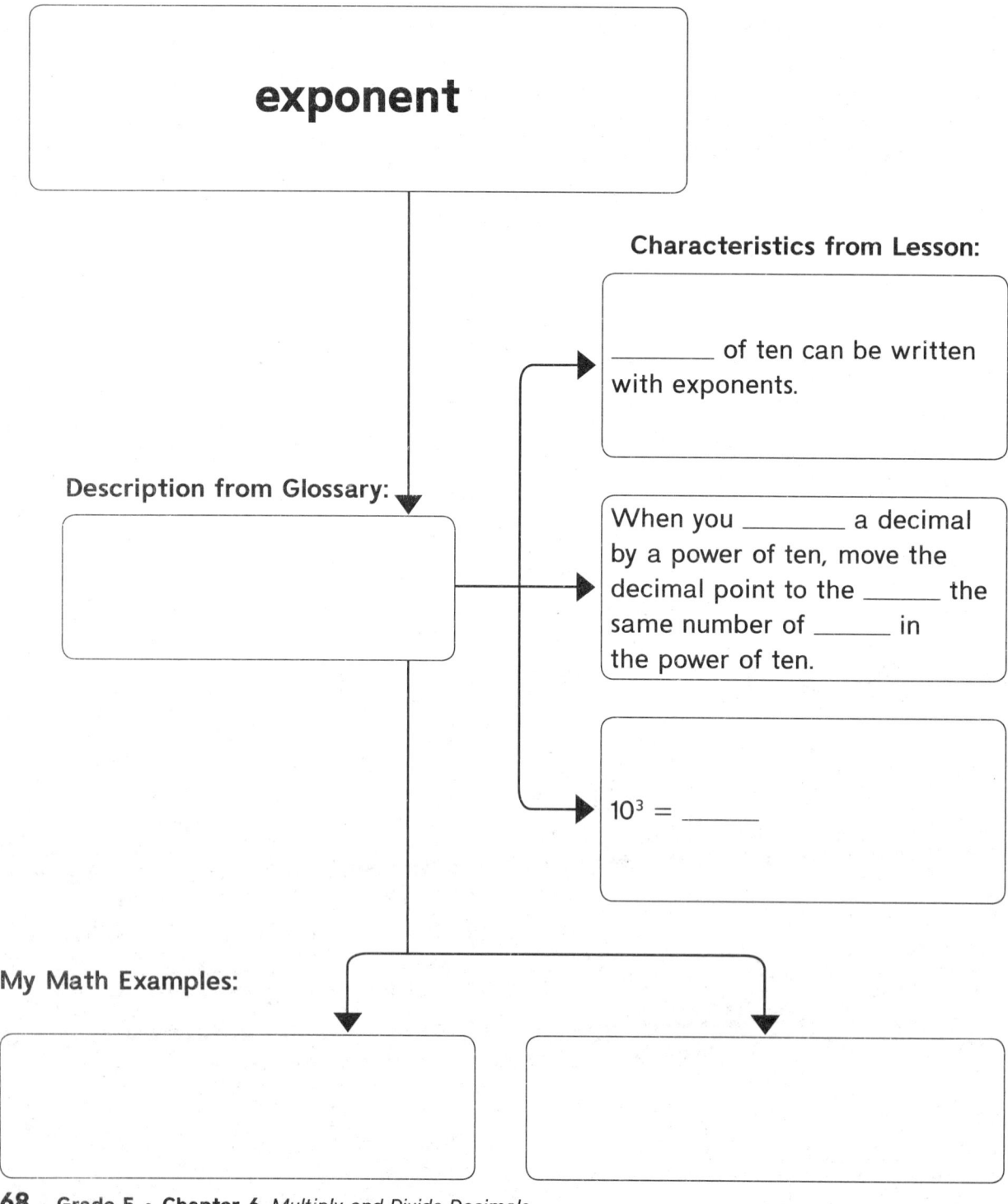

My Math Vocabulary:

**exponent**

Characteristics from Lesson:

_____ of ten can be written with exponents.

When you _____ a decimal by a power of ten, move the decimal point to the _____ the same number of _____ in the power of ten.

$10^3 = $ _____

Description from Glossary:

My Math Examples:

68   Grade 5 • Chapter 6 *Multiply and Divide Decimals*

NAME _____   DATE _____

# Chapter 7 Expressions and Patterns

*Inquiry of the Essential Question:*

**How are patterns used to solve problems?**

Read the Essential Question. Describe your observations (I see...), inferences (I think...), and prior knowledge (I know...) of each math example. Write additional questions you have below. Then share your ideas and questions with a classmate.

---

**Phrase:** subtract 2 from 8, then divide by 3

**Expression:** $(8 - 2) \div 3$

I see ...

I think...

I know...

---

72, 67, 62, 57, 52, 47, _?_

−5  −5  −5  −5  −5

The next number in the pattern is 47 − 5, or _____ .

I see ...

I think...

I know...

---

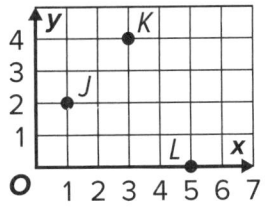

J (1, 2)
K (3, 4)
L (5, 0)

I see ...

I think...

I know...

---

Questions I have...

_____

_____

_____

NAME _____  DATE _____

# Lesson 1 Vocabulary Cognates

*Inquiry/Hands On: Numerical Expressions*

Use the Glossary to define the math word in English and in Spanish in the word boxes. Write a sentence using your math word.

| **evaluate** | **evaluar** |
|---|---|
| Definition | Definición |

My math word sentence:

| **numerical expression** | **expressión numérica** |
|---|---|
| Definition | Definición |

My math word sentence:

NAME _____ DATE _____

# Lesson 2 Vocabulary Definition Map
*Order of Operations*

Use the definition map to write a description and list characteristics about the vocabulary word or phrase. Write or draw math examples. Share your examples with a classmate.

My Math Vocabulary:

## order of operations

**Characteristics from Lesson:**

The first rule is to perform operations in _____. Then perform operations inside brackets, and finally inside _____.

The _____ rule is to find the value of _____.

The ____ rule is to _____ and _____ in order from left to right.

The _____ rule is to ____ and _____ in order from left to right.

**Description from Glossary:**

**My Math Examples:**

Grade 5 • Chapter 7 *Expressions and Patterns* **71**

NAME _____ DATE _____

# Lesson 3 Note Taking

## Write Numerical Expressions

Read the question. Write words you need help with and research each word. Use your lesson to write your Cornell notes. Write or draw math examples to explain your thinking. Share your examples with a classmate.

**Building on the Essential Question**

How do you write numerical expressions?

**Words I need help with:**

**Notes:**

A numerical expression is a combination of numbers and _____.

The numerical expression ____ + ____ represents the phrase *add three and four*. When you evaluate the expression, you get ____.

The numerical expression ____ × ____ represents the phrase *multiply seven by two*. When you evaluate the expression, you get ____.

The order of operations is a set of rules to follow when more than one _____ is used in an expression.

1. Perform operations in _____.

2. Find the value of _____.

3. _____ and _____ in order from left to right.

4. ____ and _____ in order from left to right.

The numerical expression (____ + ____) × ____ represents the phrase *add three and four, then multiply by two*.

The first operation to perform in this expression is _____ because it is enclosed in the _____.

The second operation to perform in this expression is _____.

When you evaluate the expression, you get ____.

**My Math Examples:**

72 Grade 5 • Chapter 7 *Expressions and Patterns*

NAME _____ DATE _____

# Lesson 4 Problem-Solving Investigation

## STRATEGY: Work Backward

Work backward to solve each problem.

1. **Seth** bought a movie ticket, popcorn, and a drink. After the movie, **he** played 4 video **games** that **each cost** the **same**. He spent a **total** of **$19**. How much did it **cost** to play **each** video **game**?

| Movie | Costs |
|---|---|
| Popcorn | $4 |
| Drink | $3 |
| Ticket | $8 |

| Understand | Solve |
|---|---|
| I know: | |
| I need to find: | |

| Plan | Check |
|---|---|
| Total spent: $19 | |
| I need to work _____ to solve. | |

2. Students sold raffle tickets to raise money for a field trip. The first **20 tickets** sold cost **$4 each**. To sell more tickets, they **lowered** the **price** to **$2 each**. If they **raised $216**, how many **tickets** did they **sell in all**?

| Understand | Solve |
|---|---|
| I know: | |
| I need to find: | |

| Plan | Check |
|---|---|
| Total raised: $216 | |
| I need to _____ backward to solve. | |

Grade 5 • Chapter 7 *Expressions and Patterns* 73

NAME _____  DATE _____

# Lesson 5 Guided Writing

## Inquiry/Hands On: Generate Patterns

**How do you identify a generated pattern?**

Use the exercises below to help you build on answering the Essential Question. Write the correct word or phrase on the lines provided.

1. Rewrite the question in your own words.
   _____
   _____

2. What key words do you see in the question?
   _____

3. The shape below is made with ____ lines.

4. The shape below is made with ____ lines. Which is ____ more lines than the previous shape.

5. The shape below is made with ____ lines. Which is ____ more lines than the previous shape.

6. If the pattern continues, the next shape will have ____ more lines than the previous shape. The next shape in the pattern will be made with ____ lines.

7. The shape below is made with ____ lines.

8. How do you identify a pattern?
   _____
   _____
   _____
   _____

74  Grade 5 • Chapter 7 *Expressions and Patterns*

NAME _____ DATE _____

# Lesson 6 Concept Web

*Patterns*

Write the next term in each sequence on the concept web.

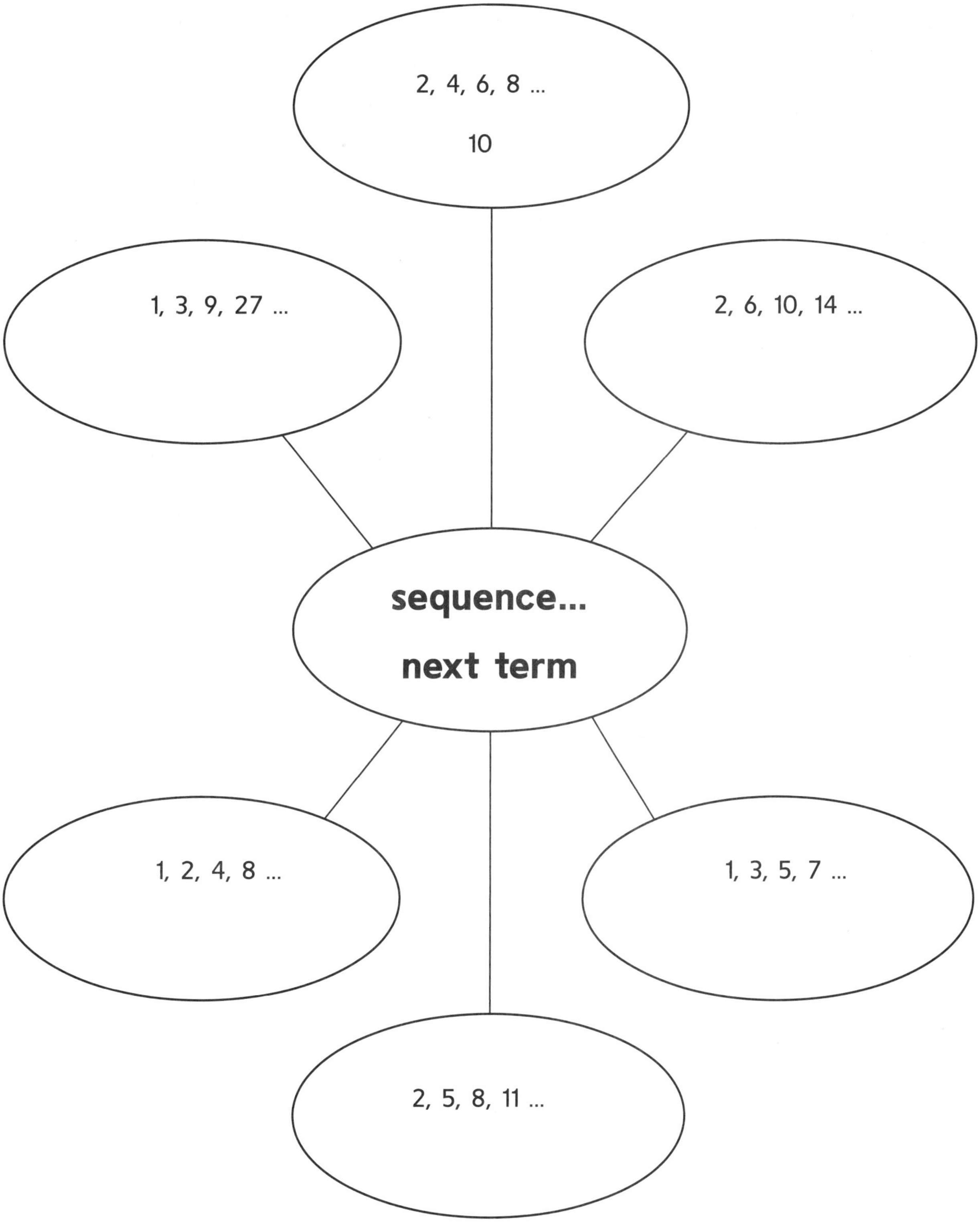

- 2, 4, 6, 8 …
  10

- 1, 3, 9, 27 …

- 2, 6, 10, 14 …

**sequence… next term**

- 1, 2, 4, 8 …

- 1, 3, 5, 7 …

- 2, 5, 8, 11 …

Grade 5 • Chapter 7 *Expressions and Patterns* **75**

NAME _____  DATE _____

# Lesson 7 Note Taking

## Inquiry/Hands On: Map Locations

Read the question. Write words you need help with and research each word. Use your lesson to write your Cornell notes. Write or draw math examples to explain your thinking. Share your examples with a classmate.

| Building on the Essential Question | Notes: |
|---|---|
| How do you plot the location of an item on a map? | This grid represents locations on a map. |
| | When you move ____ on the grid you move **north** on the map. |
| | When you move ____ on the grid you move **south** on the map. |
| **Words I need help with:** | When you move ____ on the grid you move **west** on the map. |
| | When you move ____ on the grid you move **east** on the map. |
| | Each square on the grid represents a **block** on the map. |
| | The library is _____ blocks _____ of the school. |
| | The park is ____ blocks ____ of the school. |
| | Marcia's home is one block south of the park. |
| | The dot that represents Marcia's home will be ____ square _____ the dot that represents the park. Draw and label a dot for Marcia's home. |

**My Math Examples:**

76  Grade 5 • Chapter 7 *Expressions and Patterns*

NAME _____ DATE _____

# Lesson 8 Vocabulary Chart
*Ordered Pairs*

Use the three-column chart to organize the vocabulary in this lesson. Write the word in Spanish. Then write the correct terms to complete each definition.

| English | Spanish | Definition |
|---|---|---|
| **coordinate** | | One of two numbers in an _____ pair. |
| **coordinate plane** | | A plane that is formed when ____ number lines _____. |
| **ordered pair** | | A ____ of numbers that is used to name a ____ on the coordinate plane. |
| **origin** | | The point (0, 0) on a coordinate plane where the _____ axis meets the _____ axis. |
| ***x*-coordinate** | | The ____ part of an ordered pair that indicates how far to the ____ of the *y*-axis the corresponding point is. |
| ***y*-coordinate** | | The _____ part of an ordered pair that indicates how far _____ the *x*-axis the corresponding point is. |

Grade 5 • Chapter 7 *Expressions and Patterns* 77

NAME _____ DATE _____

# Lesson 9 Multiple Meaning Word
*Graph Patterns*

Complete the four-square chart to review the multiple meaning word or phrase.

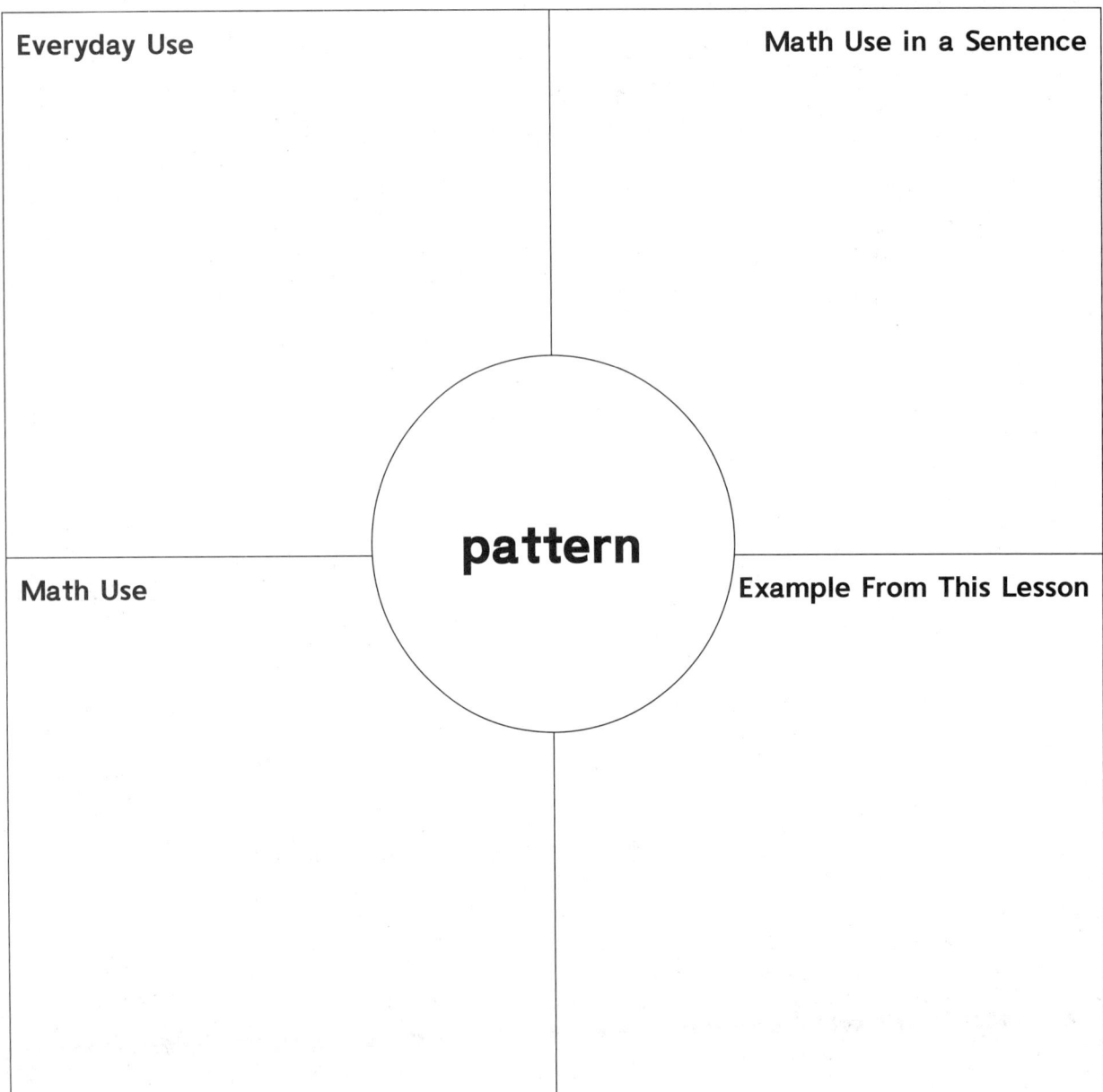

Write the correct term on the line to complete the sentence.

When you graph a pattern of _____, the points are plotted from the bottom left to the top right on the graph.

78   Grade 5 • Chapter 7 *Expressions and Patterns*

NAME _____  DATE _____

# Chapter 8 Fractions and Decimals

*Inquiry of the Essential Question:*

**How are factors and multiples helpful in solving problems?**

Read the Essential Question. Describe your observations (I see..), inferences (I think...), and prior knowledge (I know...) of each math example. Write additional questions you have below. Then share your ideas and questions with a classmate.

$$\frac{5}{9} = 5 \div 9$$

I see ...

I think...

I know...

---

Two pounds of grapes are divided equally among five fruit baskets. How many pounds of grapes are in each basket?

| 1 | 2 | 3 | 4 | 5 | | 1 | 2 | 3 | 4 | 5 |

Each basket has $\frac{2}{5}$ pounds of grapes.

I see ...

I think...

I know...

---

$\frac{1}{5} = \frac{1 \times 2}{5 \times 2} = \frac{2}{10}$

Write the fraction with a denominator of 10.

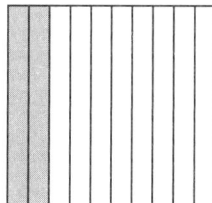

I see ...

I think...

I know...

---

Questions I have...

_____

_____

_____

Grade 5 • Chapter 8 *Fractions and Decimals* **79**

NAME _____ DATE _____

# Lesson 1 Vocabulary Cognates

*Fractions and Division*

Use the Glossary to define the math word in English and in Spanish in the word boxes. Write a sentence using your math word.

| fraction | fracción |
|---|---|
| Definition | Definición |
| My math word sentence: | |

| numerator | numerador |
|---|---|
| Definition | Definición |
| My math word sentence: | |

| denominator | denominador |
|---|---|
| Definition | Definición |
| My math word sentence: | |

80  Grade 5 • Chapter 8 *Fractions and Decimals*

# Lesson 2 Vocabulary Definition Map
*Greatest Common Factor*

Use the definition map to write a description and list characteristics about the vocabulary word or phrase. Write or draw math examples. Share your examples with a classmate.

My Math Vocabulary:

**greatest common factor**

**Description from Glossary:**

**Characteristics from Lesson:**

A _____ is a number that is multiplied by another number.

A common factor is a number that is a factor of _____ or more numbers.

Prime factorization is a way of expressing a _____ number as a product of its _____ factors.

**My Math Examples:**

Grade 5 • Chapter 8 *Fractions and Decimals* **81**

NAME _____ DATE _____

# Lesson 3 Note Taking

## Simplest Form

Read the question. Write words you need help with and research each word. Use your lesson to write your Cornell notes. Write or draw math examples to explain your thinking. Share your examples with a classmate.

| Building on the Essential Question | Notes: |
|---|---|
| How do you write a fraction in simplest form? | $\frac{9}{12}$ is the fraction to simplify.<br>The numerator of the fraction is ____ and the denominator is ____.<br>The factors of 9 are ____, ____, ____.<br>The factors of 12 are ____, ____, ____, ____, ____, ____.<br>The common factors of 9 and 12 are ____ and ____.<br>The **greatest** common factor (GCF) of 9 and 12 is ____.<br>Divide both the numerator and denominator by the GCF.<br>9 ÷ ____ = ____<br>12 ÷ ____ = ____<br>A fraction is in _____ form when the greatest common factor (GCF) of the numerator and the denominator is 1.<br>The factors of 3 are ____ and ____.<br>The factors of 4 are ____, ____, ____.<br>The common factor of 3 and 4 is ____.<br>The greatest common factor (GCF) of 3 and 4 is ____.<br>Equivalent fractions are fractions that have the _____ value.<br>$\frac{9}{12}$ and — are equivalent fractions. |
| Words I need help with: | |

**My Math Examples:**

82 Grade 5 • Chapter 8 *Fractions and Decimals*

NAME _____ DATE _____

# Lesson 4 Problem-Solving Investigation
## STRATEGY: Guess, Check, and Revise

Guess, check, and revise to solve each problem.

1. Bike path A is **4 miles** long.
   Bike path B is **7 miles** long.
   If **April** biked a **total** of **37 miles**,
   how many **times** did **she** bike **each** path?

| Understand<br>I know:<br><br>I need to find: | Solve |
|---|---|
| Plan<br><br>I will _____, check, and revise to solve. | Check |

2. Ruben sees **14 wheels** on a total of **6** bicycles **and** tricycles. How **many** bicycles **and** tricycles are there?

| Understand<br>I know:<br><br>I need to find: | Solve |
|---|---|
| Plan<br><br>A bicycle has _____ wheels.<br>A tricycle has _____ wheels.<br><br>I will guess, _____, and revise to solve. | Check |

Grade 5 • Chapter 8 *Fractions and Decimals* **83**

NAME _____ DATE _____

# Lesson 5 Vocabulary Chart

*Least Common Multiple*

Use the three-column chart to organize the vocabulary in this lesson. Write the word in Spanish. Then write the correct terms to complete each definition.

| English | Spanish | Definition |
|---|---|---|
| **multiple** | | A multiple of a number is the product of _____ number and any _____ number. |
| **common multiple** | | A _____ number that is a _____ of two or more numbers. |
| **least common multiple (LCM)** | | The _____ whole number greater than 0 that is a _____ multiple of each of two or more numbers. |
| **product** | | The _____ to a multiplication problem. |

84 Grade 5 • Chapter 8 *Fractions and Decimals*

# Lesson 6 Concept Web

*Compare Fractions*

Use the concept web to identify the least common multiple of each set of numbers.

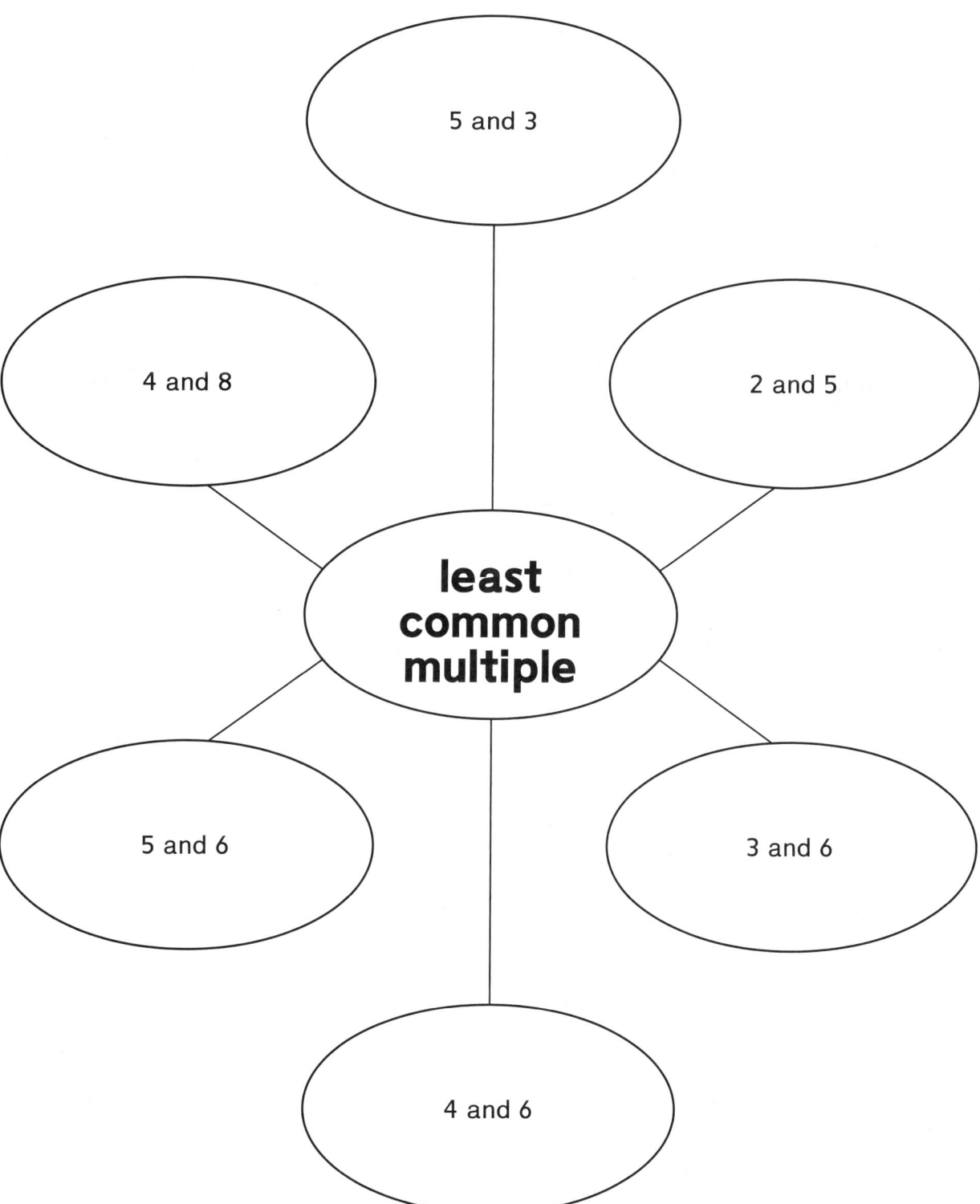

Grade 5 • Chapter 8 *Fractions and Decimals*

NAME _____  DATE _____

# Lesson 7 Guided Writing

## Inquiry/Hands On: Use Models to Write Fractions as Decimals

**How do you use models to write fractions as decimals?**

Use the exercises below to help you build on answering the Essential Question. Write the correct word or phrase on the lines provided.

1. Rewrite the question in your own words.
   _____
   _____

2. What key words do you see in the question?
   _____

3. A decimal can be modeled by shaded squares on a 10 by 10 grid. The grid contains _____ squares.

   One shaded square represents _____ .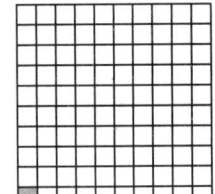

   Ten shaded squares represents _____ .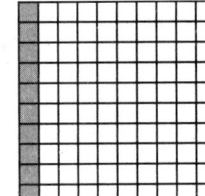

4. When modeling a fraction on a 10 by 10 grid, the first step is to find an _____ fraction with a denominator of 10 or 100.

5. Equivalent fractions are fractions that have the _____ value. If you multiply the _____ and _____ by the same number, you will find an equivalent fraction.

6. The following fractions are equivalent: $\frac{2}{5} = \frac{}{100}$.

7. How many squares will be shaded to represent this equivalent fraction? ____

8. The fraction $\frac{}{100}$ and the decimal _____ is modeled on the grid. The fraction $\frac{2}{5}$ written as a decimal is _____.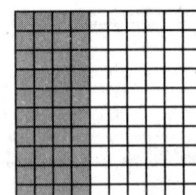

9. How do you use models to write fractions as decimals?
   _____
   _____

NAME _____ DATE _____

# Lesson 8 Vocabulary Cognates

*Write Fractions as Decimals*

Use the Glossary to define the math word in English and in Spanish in the word boxes. Write a sentence using your math word.

| **equivalent fractions** | **fracciones equivalentes** |
|---|---|
| Definition | Definición |

My math word sentence:

| **equivalent decimals** | **decimales equivalentes** |
|---|---|
| Definition | Definición |

My math word sentence:

Grade 5 • Chapter 8 *Fractions and Decimals*  87

NAME _____ DATE _____

# Chapter 9 Add and Subtract Fractions

*Inquiry of the Essential Question:*

How can equivalent fractions help me add and subtraction fractions?

Read the Essential Question. Describe your observations (I see...), inferences (I think...), and prior knowledge (I know...) of each math example. Write additional questions you have below. Then share your ideas and questions with a classmate.

Use models to show sums like $\frac{3}{5} + \frac{4}{5}$.

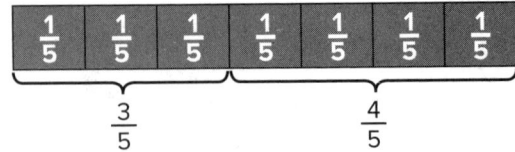

There are seven $\frac{1}{5}$-tiles. So, the sum is $\frac{7}{5}$ or $1\frac{2}{5}$.

I see ...

I think...

I know...

---

$\frac{7}{9} - \frac{4}{9} = \frac{7-4}{9}$

$= \frac{3}{9}$ or $\frac{1}{3}$

I see ...

I think...

I know...

---

Solve subtraction problems like $9\frac{4}{5} - 1\frac{3}{10}$.

$9\frac{4}{5} \rightarrow 9\frac{8}{10}$   Write $9\frac{4}{5}$ as $9\frac{8}{10}$.

$-1\frac{3}{10} \rightarrow -1\frac{3}{10}$

$8\frac{5}{10}$ or $8\frac{1}{2}$

I see ...

I think...

I know...

---

Questions I have...

_____

_____

_____

88 Grade 5 • Chapter 9 *Add and Subtract Fractions*

NAME _____ DATE _____

# Lesson 1 Vocabulary Definition Map
*Round Fractions*

Use the definition map to write a description and list characteristics about the vocabulary word or phrase. Write or draw math examples. Share your examples with a classmate.

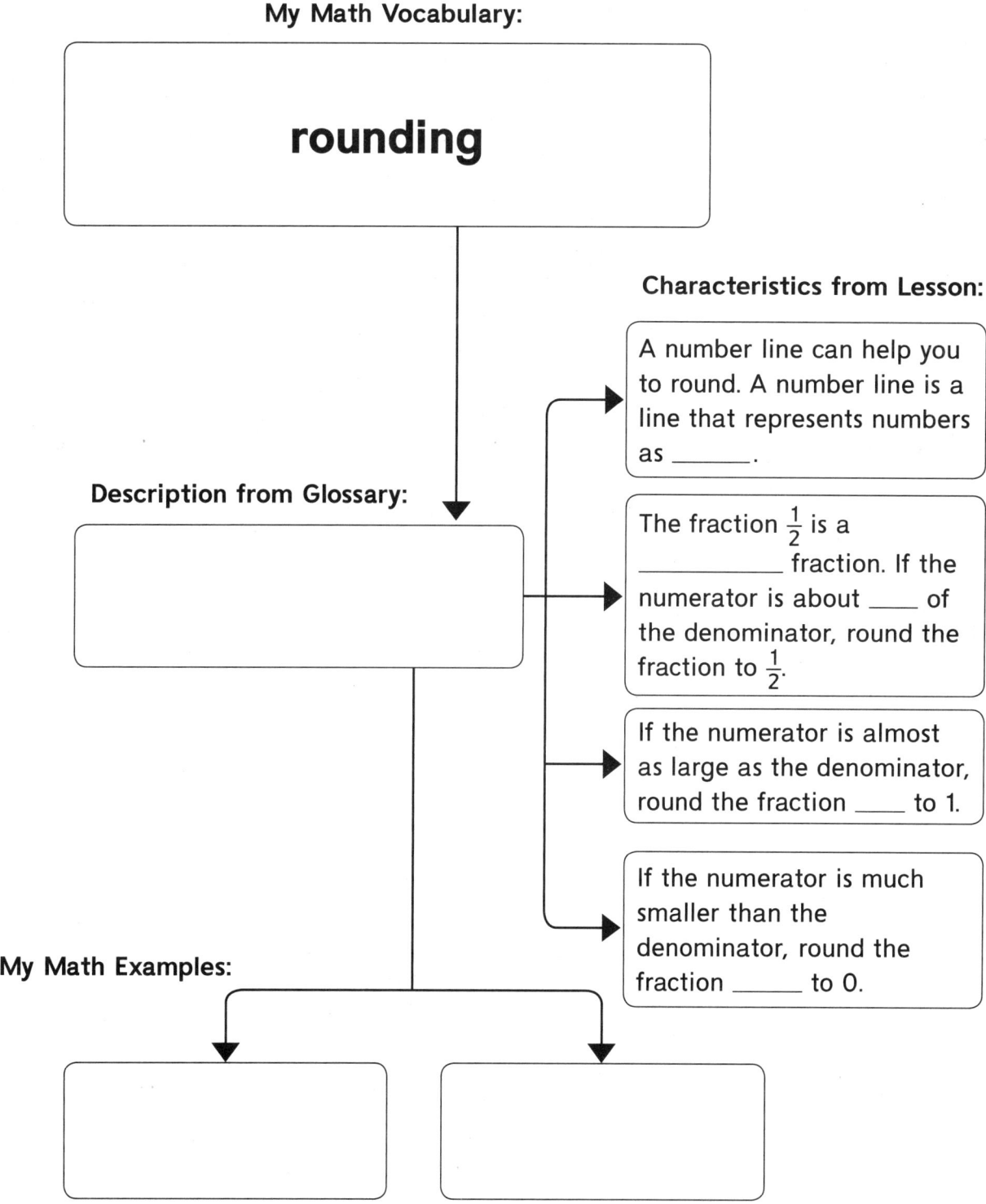

My Math Vocabulary:

**rounding**

**Characteristics from Lesson:**

A number line can help you to round. A number line is a line that represents numbers as _____.

The fraction $\frac{1}{2}$ is a _____ fraction. If the numerator is about ____ of the denominator, round the fraction to $\frac{1}{2}$.

If the numerator is almost as large as the denominator, round the fraction ____ to 1.

If the numerator is much smaller than the denominator, round the fraction ____ to 0.

**Description from Glossary:**

**My Math Examples:**

Grade 5 • Chapter 9 *Add and Subtract Fractions* **89**

# Lesson 2 Multiple Meaning Word

*Add Like Fractions*

Complete the four-square chart to review the multiple meaning word.

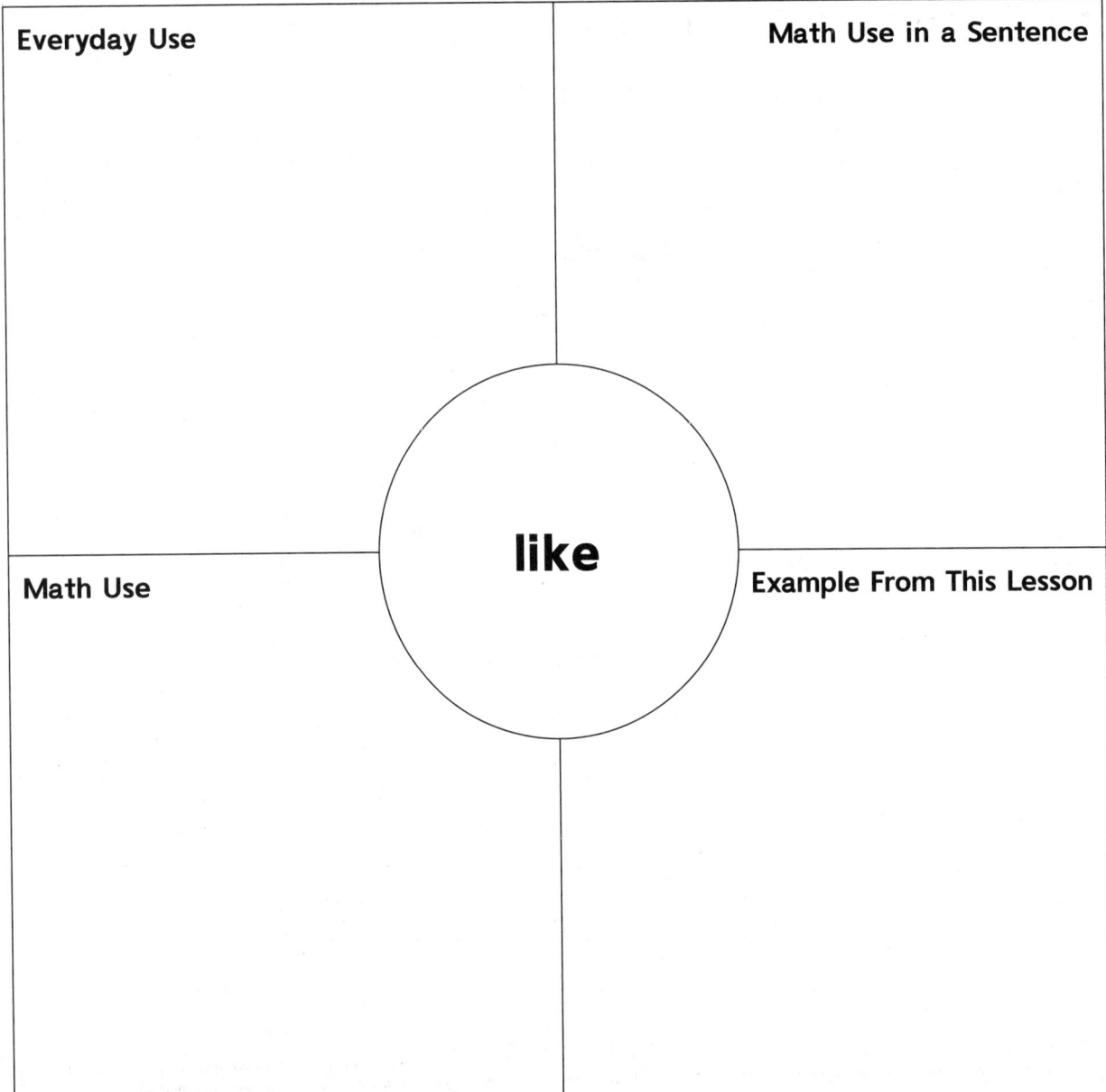

Write the correct terms on the blank lines to complete the sentence.

To add like fractions, you add the _____ and keep the denominator the _____.

# Lesson 3 Concept Web

*Subtract Like Fractions*

Use the concept web to identify examples of like fractions. Write *true* or *false*.

- $\frac{1}{3}$ and $\frac{2}{3}$
- $\frac{2}{5}$ and $\frac{2}{6}$
- $\frac{4}{6}$ and $\frac{5}{6}$

**These are like fractions. True or False?**

- $\frac{7}{8}$ and $\frac{5}{8}$
- $\frac{1}{4}$ and $\frac{3}{4}$
- $\frac{2}{3}$ and $\frac{3}{4}$

Grade 5 • Chapter 9 Add and Subtract Fractions

NAME _____ DATE _____

# Lesson 4 Note Taking

*Inquiry/Hands On: Use Models to Add Unlike Fractions*

Read the question. Write words you need help with and research each word. Use your lesson to write your Cornell notes. Write or draw math examples to explain your thinking. Share your examples with a classmate.

| Building on the Essential Question | Notes: |
|---|---|
| How do you use models to add unlike fractions? | The fraction modeled below is $\frac{2}{3}$. The denominator is ____. |
| | $\boxed{\frac{1}{3} \mid \frac{1}{3}}$ |
| | Fractions that have _____ denominators are called unlike fractions. |
| | The fraction $\frac{1}{6}$ and the fraction $\frac{2}{3}$ are _____ fractions. |
| | To add unlike fractions, the fractions must have a _____ denominator. |
| **Words I need help with:** | ____ $\frac{1}{6}$-tiles will match the length of $\frac{1}{3}$. |
| | ____ $\frac{1}{6}$-tiles will match the length of $\frac{2}{3}$. |
| | $\boxed{\frac{1}{6} \mid \frac{1}{6} \mid \frac{1}{6} \mid \frac{1}{6} \mid \frac{1}{6}}$ |
| | $\boxed{\frac{1}{3} \mid \frac{1}{3}}$ |
| | $\frac{1}{6} + \frac{4}{6} = \frac{}{6}$  So, $\frac{1}{6} + \frac{2}{3} = \frac{}{6}$ |
| | There are __ $\frac{1}{6}$-tiles altogether. |

**My Math Examples:**

NAME _____ DATE _____

# Lesson 5 Vocabulary Cognates

*Add Unlike Fractions*

Use the Glossary to define the math word in English and in Spanish in the word boxes. Write a sentence using your math word.

| like fractions | fracciones semejantes |
|---|---|
| Definition | Definición |
| My math word sentence: ||

| unlike fractions | fracciones no semejantes |
|---|---|
| Definition | Definición |
| My math word sentence: ||

Grade 5 • Chapter 9 *Add and Subtract Fractions*

# Lesson 6 Guided Writing

*Inquiry/Hands On: Use Models to Subtract Unlike Fractions*

**How do you subtract unlike fractions using models?**

Use the exercises below to help you build on answering the Essential Question. Write the correct word or phrase on the lines provided.

1. Rewrite the question in your own words.
   _____
   _____

2. What key words do you see in the question?
   _____

3. The fractions modeled right are – and –.

   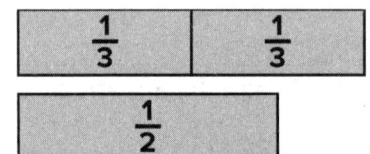

4. The length of the fraction tile for $\frac{1}{2}$ is ____ than the fraction tiles for $\frac{2}{3}$.

5. The subtraction expression – – – is modeled below.

   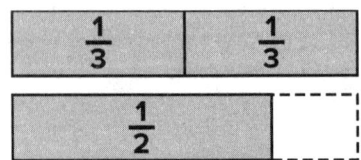

6. You can find the _____ of a modeled subtraction expression by finding which fraction tiles will ____ ____ the area of the dashed box.

7. Which of the following tiles will fill in the dashed box: a $\frac{1}{2}$-tile, a $\frac{1}{3}$-tile, or a $\frac{1}{6}$-tile?
   _____

8. Find the difference. $\frac{2}{3} - \frac{1}{2} = -$

9. How do you subtract unlike fractions using models?
   _____
   _____
   _____

94 Grade 5 • Chapter 9 *Add and Subtract Fractions*

NAME _____ DATE _____

# Lesson 7 Concept Web

## Subtract Unlike Fractions

Use the concept web to identify the least common denominator used to subtract unlike fractions.

- $\frac{2}{3}$ and $\frac{1}{2}$
- $\frac{4}{5}$ and $\frac{1}{4}$
- $\frac{1}{3}$ and $\frac{3}{4}$

**least common denominator**

- $\frac{2}{5}$ and $\frac{1}{2}$
- $\frac{1}{3}$ and $\frac{3}{5}$
- $\frac{3}{4}$ and $\frac{1}{2}$

Grade 5 · Chapter 9 Add and Subtract Fractions  **95**

# Lesson 8 Problem-Solving Investigation

## STRATEGY: Determine Reasonable Answers

Solve each problem by determining a reasonable answer.

1. Use the **table** to determine whether **245** pounds, **260** pounds, or **263** pounds is the most **reasonable estimate** for how much **more** the ostrich weighs **than** the flamingo. Explain.

| Bird | Weight (lb) |
|---|---|
| Flamingo | $9\frac{1}{10}$ |
| Ostrich | $253\frac{1}{2}$ |

| Understand | Solve |
|---|---|
| I know: <br><br> I need to find: | |
| **Plan** <br> 1. Round the weight of a flamingo. <br> 2. Round the weight of an ostrich. <br> 3. Find the difference of the rounded weights. | **Check** <br> My answer is reasonable because… |

2. A grocer sells **12** pounds of apples. Of those, $5\frac{3}{4}$ pounds are **green** and $3\frac{1}{4}$ pounds are **golden**. The **rest** are **red**. Which is a more **reasonable** estimate for how many pounds of **red apples** the grocer **sold**? **3** pounds or **5** pounds? Explain.

| Understand | Solve |
|---|---|
| I know: <br><br> I need to find: | |
| **Plan** <br> 1. Round the weight of green apples sold. <br> 2. Round the weight of golden apples sold. <br> 3. Subtract the rounded weight of green apples sold. <br> 4. Subtract the rounded weight of golden apples sold. | **Check** <br> My answer is reasonable because… |

NAME _____ DATE _____

# Lesson 9 Vocabulary Chart

*Estimate Sums and Differences*

Use the three-column chart to organize the review vocabulary in this lesson. Write the word in Spanish. Then write the correct terms to complete each definition.

| English | Spanish | Definition |
|---|---|---|
| **estimate** | | A number close to an _____ value. An _____ indicates about how much. |
| **sum** | | The answer to an _____ problem. |
| **fraction** | | A number that represents part of a _____ or part of a ____. |
| **like fractions** | | Fractions that have the _____ denominator. |
| **unlike fractions** | | Fractions that have _____ denominators. |

Grade 5 • Chapter 9 Add and Subtract Fractions   **97**

NAME _____ DATE _____

# Lesson 10 Note Taking

## Inquiry/Hands On: Use Models to Add Mixed Numbers

Read the question. Write words you need help with and research each word. Use your lesson to write your Cornell notes. Write or draw math examples to explain your thinking. Share your examples with a classmate.

**Building on the Essential Question**

How do you use models to add mixed numbers?

**Words I need help with:**

**Notes:**

The mixed number modeled below is $\_\frac{\_}{\_}$.

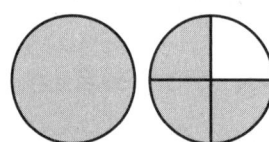

The mixed number modeled below is $\_\frac{\_}{\_}$.

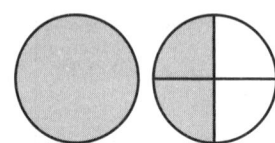

To find the sum of two mixed numbers, _____ the models of the fractions.

A whole fraction circle is equal to __ quarter fraction circles.

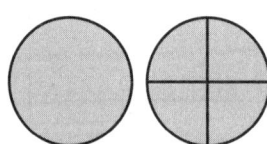

Find the sum of $1\frac{3}{4} + 1\frac{2}{4}$.

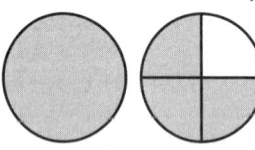 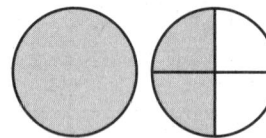

There are a total of __ whole fraction circles.

There are a total of __ quarter fraction circles.

$1\frac{3}{4} + 1\frac{2}{4} = 1 + \frac{1}{4} + \frac{1}{4} + \frac{1}{4} + 1 + \frac{1}{4} + \frac{1}{4} = \_ + \frac{\_}{4}$

$= \_ + \frac{\_}{4} = \frac{\_}{4}$

$1\frac{3}{4} + 1\frac{2}{4} = \frac{\_}{4}$

**My Math Examples:**

NAME _____ DATE _____

# Lesson 11 Vocabulary Definition Map

*Add Mixed Numbers*

Use the definition map to write a description and list characteristics about the vocabulary word or phrase. Write or draw math examples. Share your examples with a classmate.

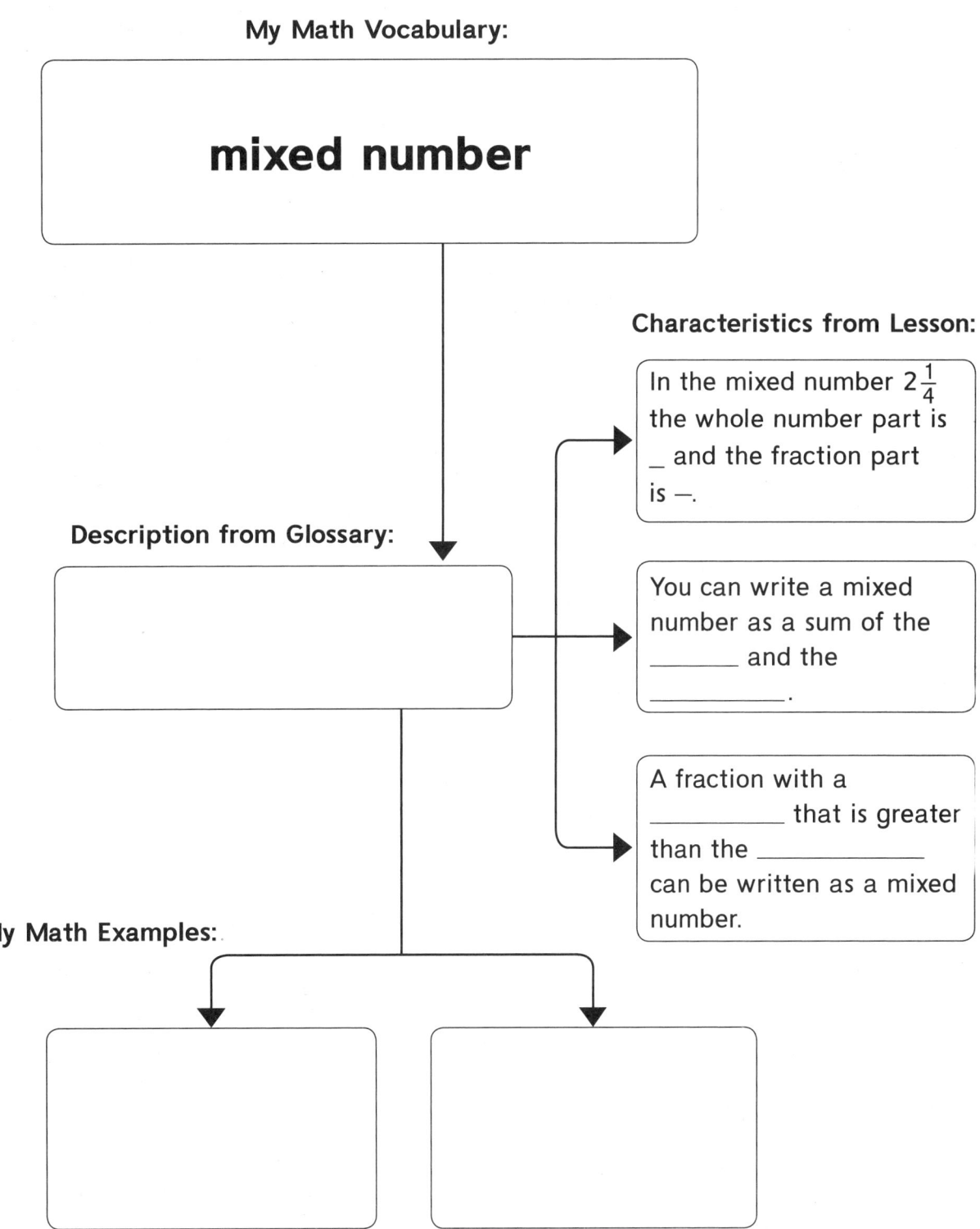

My Math Vocabulary:

**mixed number**

**Characteristics from Lesson:**

In the mixed number $2\frac{1}{4}$ the whole number part is __ and the fraction part is —.

You can write a mixed number as a sum of the _____ and the _____.

A fraction with a _____ that is greater than the _____ can be written as a mixed number.

**Description from Glossary:**

**My Math Examples:**

Grade 5 • Chapter 9 *Add and Subtract Fractions* **99**

NAME _____  DATE _____

# Lesson 12 Four-Square Vocabulary
*Subtract Mixed Numbers*

Write the definition for each math word. Write what each word means in your own words. Draw or write examples that show each math word meaning. Then write your own sentences using the words.

| Definition | My Own Words |
|---|---|
| **simplest form** | |
| My Examples | My Sentence |

| Definition | My Own Words |
|---|---|
| **equivalent fractions** | |
| My Examples | My Sentence |

100  Grade 5 • Chapter 9 *Add and Subtract Fractions*

NAME _____  DATE _____

# Lesson 13 Guided Writing

*Subtract with Renaming*

**How do you subtract mixed numbers with renaming?**

Use the exercises below to help you build on answering the Essential Question. Write the correct word or phrase on the lines provided.

1. Rewrite the question in your own words.
   _____
   _____

2. What key words do you see in the question?
   _____

3. The fractional part of the mixed number $2\frac{2}{5}$ is — and the fractional part of $1\frac{1}{2}$ is —.

   The fractions $\frac{2}{5}$ and $\frac{1}{2}$ are _____ fractions.

4. Before subtracting mixed numbers, find an equivalent fraction so the fractional parts have the _____ denominator.

5. The least common denominator for $\frac{2}{5}$ and $\frac{1}{2}$ is ___.

   $2\frac{2}{5} = 2\frac{}{10}$ and $1\frac{1}{2} = 1\frac{}{10}$

6. The mixed number $2\frac{4}{10}$ is _____ than the mixed number $1\frac{5}{10}$. The fraction $\frac{4}{10}$ is _____ than $\frac{5}{10}$.

7. Since the fractional part of the minuend ($2\frac{4}{10}$) is _____ than the fractional part of the subtrahend ($1\frac{5}{10}$), rename one whole into a fraction with a denominator of 10.

8. $2\frac{4}{10} = 1 + 1 + \frac{4}{10} = 1 + \frac{}{10} + \frac{4}{10} = 1 + \frac{}{10} = 1\frac{}{10}$

9. Now, you can subtract the mixed numbers: $2\frac{2}{5}$ and $1\frac{1}{2} = 2\frac{4}{10} - 1\frac{5}{10} = 1\frac{}{10} - 1\frac{5}{10} = \frac{}{10}$

10. How do you subtract mixed numbers with renaming?
    _____
    _____
    _____
    _____

# Chapter 10 Multiply and Divide Fractions

*Inquiry of the Essential Question:*

**What strategies can be used to multiply and divide fractions?**

Read the Essential Question. Describe your observations (I see..), inferences (I think...), and prior knowledge (I know...) of each math example. Write additional questions you have below. Then share your ideas and questions with a classmate.

Use models to find products like $\frac{2}{3} \times 21$.

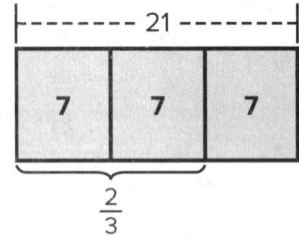

$\frac{2}{3} \times 21 = 7 + 7 = 14$

I see...

I think...

I know...

---

Use models to find products like $\frac{2}{3} \times \frac{1}{5}$.

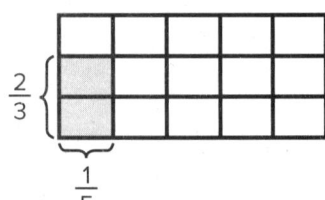

2 out of 15 sections are shaded. So, $\frac{2}{3} \times \frac{1}{5} = \frac{2}{15}$.

I see...

I think...

I know...

---

Use models to find quotients like $5 \div \frac{1}{2}$.

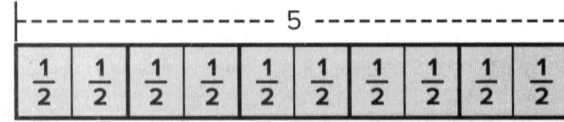

There are 10 halves in the model.
So, $5 \div \frac{1}{2} = 10$.

I see...

I think...

I know...

---

Questions I have...

_____

_____

_____

# Lesson 1 Note Taking

*Inquiry/Hands On: Part of a Number*

Read the question. Write words you need help with and research each word. Use your lesson to write your Cornell notes. Write or draw math examples to explain your thinking. Share your examples with a classmate.

**Building on the Essential Question**

How can you use parts of a number to multiply and divide?

**Words I need help with:**

**Notes:**

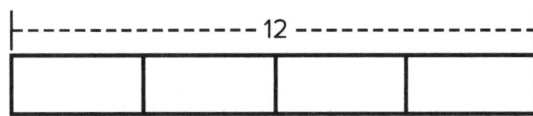

The bar diagram represents the number ____.

The bar diagram is divided into ____ equal sections.

Each section of the bar represents $\frac{1}{\phantom{0}}$ of the whole.

$12 \div 4$ is the same as $\frac{1}{\phantom{0}}$ of 12

$12 \div 4$ is the same as $\frac{1}{\phantom{0}} \times 12$

$\frac{1}{\phantom{0}}$ of 12 is the same as $\frac{1}{\phantom{0}} \times 12$

Each section of the bar represents the number ____.

$12 \div 4 =$ ____

$\frac{1}{4} \times 12 =$ ____

$\frac{1}{4}$ of 12 = ____

**My Math Examples:**

Grade 5 • Chapter 10 *Multiply and Divide Fractions* 103

NAME _____ DATE _____

# Lesson 2 Four-Square Vocabulary
*Estimate Products of Fractions*

Complete the four-square chart to review the multiple meaning word or phrase.

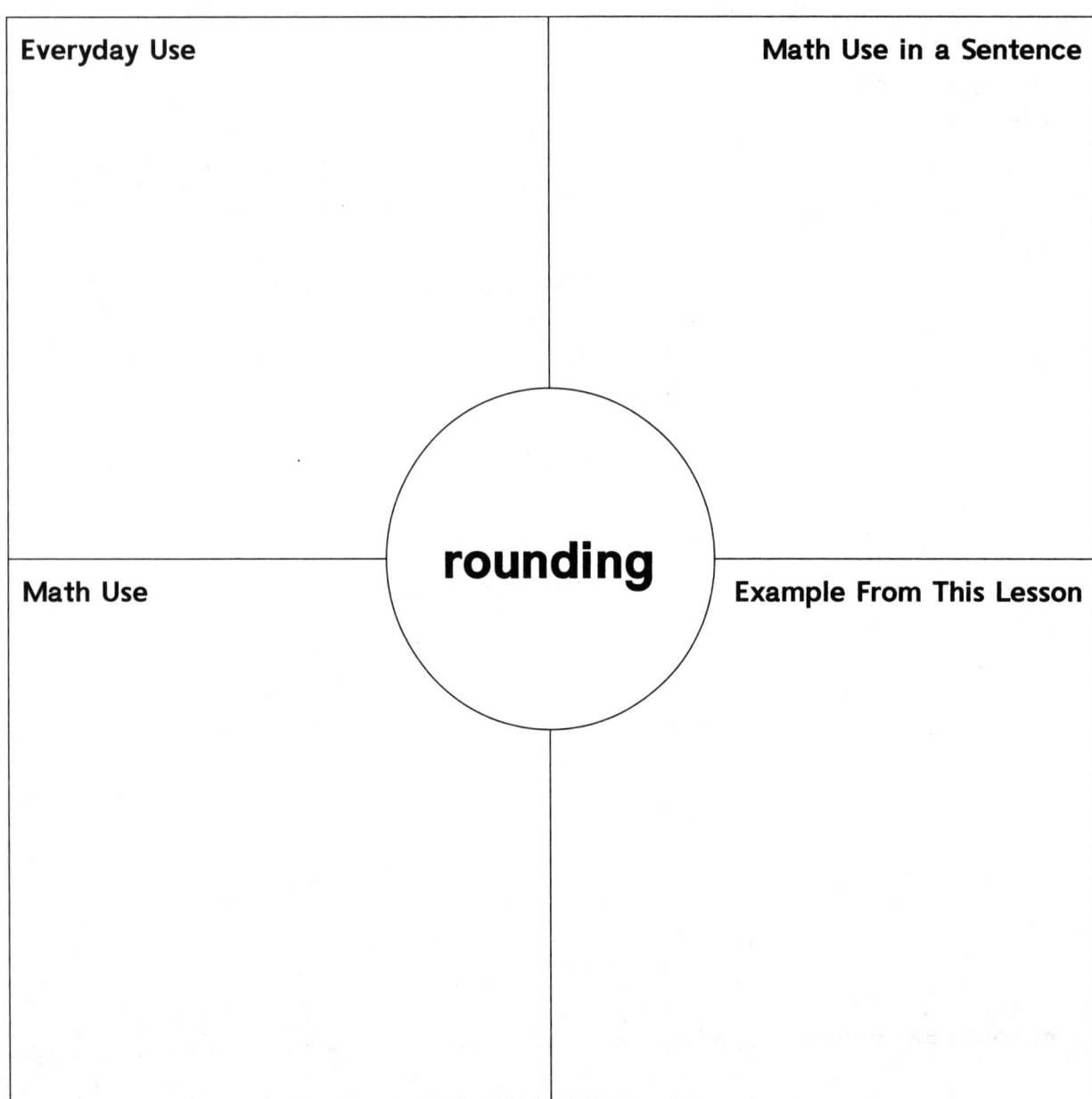

Write the correct terms on the blank lines to complete the sentence.

You can _____ products of fractions using rounding and _____ numbers.

104 Grade 5 • Chapter 10 *Multiply and Divide Fractions*

# Lesson 3 Guided Writing

## Inquiry/Hands On: Model Fraction Multiplication

**How do you model fraction multiplication?**

Use the exercises below to help you build on answering the Essential Question. Write the correct word or phrase on the lines provided.

1. Rewrite the question in your own words.
   _____
   _____

2. What key words do you see in the question?
   _____

3. Multiplication can be represented as _____ addition.

4. ___ × $\frac{1}{4}$ = $\frac{\phantom{0}}{\phantom{0}}$ + $\frac{\phantom{0}}{\phantom{0}}$ + $\frac{\phantom{0}}{\phantom{0}}$

5. To model the fraction $\frac{1}{4}$, the model will have ___ equal sections with ___ section(s) shaded.

6. The following repeated addition expression is modeled below: $\frac{\phantom{0}}{\phantom{0}}$ + $\frac{\phantom{0}}{\phantom{0}}$ + $\frac{\phantom{0}}{\phantom{0}}$

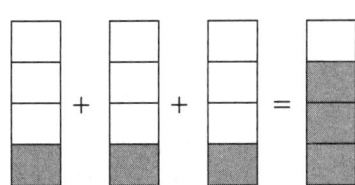

7. There are ___ total sections shaded in the repeated addition expression. The sum will have ___ sections shaded. The numerator in the sum will be ___. The denominator will be the same, which is ___.

8. $\frac{1}{4} + \frac{1}{4} + \frac{1}{4}$ = $\frac{\phantom{0}}{\phantom{0}}$

9. 3 × $\frac{1}{4}$ = $\frac{\phantom{0}}{\phantom{0}}$

10. How do you model fraction multiplication?
    _____
    _____
    _____

**Grade 5 • Chapter 10** *Multiply and Divide Fractions*

NAME _____ DATE _____

# Lesson 4 Vocabulary Cognates

*Multiply Whole Numbers and Fractions*

Use the Glossary to define the math word in English and in Spanish in the word boxes. Write a sentence using your math word.

| **Commutative Property** | **propiedad conmutativa** |
|---|---|
| Definition | Definición |
| My math word sentence: ||

| **unknown** | **incógnita** |
|---|---|
| Definition | Definición |
| My math word sentence: ||

# Lesson 5 Note Taking

## Inquiry/Hands On: Use Models to Multiply Fractions

Read the question. Write words you need help with and research each word. Use your lesson to write your Cornell notes. Write or draw math examples to explain your thinking. Share your examples with a classmate.

| | |
|---|---|
| **Building on the Essential Question**<br><br>How do you use models to multiply fractions? | **Notes:**<br><br>The fraction modeled below is —.<br><br><br>The fraction modeled below is —.<br>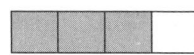<br><br>To model the multiplication of these two fractions, divide a square into ____ equal rows since the denominator of the first fraction is 2. Divide the square into ____ equal columns since the denominator of the second fraction is 4.<br><br>The multiplication expression below is — × —.<br>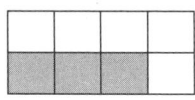<br><br>The shaded portion is where — and — **intersect**.<br><br>There are ____ sections **shaded** in the multiplication model. This represents the _____ in the product.<br><br>There are ____ **total** sections in the multiplication model. This represents the _____ in the product.<br><br>$\frac{1}{2} \times \frac{3}{4} = -$ |
| **Words I need help with:** | |
| **My Math Examples:** | |

Grade 5 • Chapter 10 Multiply and Divide Fractions

NAME _____ DATE _____

# Lesson 6 Vocabulary Chart
*Multiply Fractions*

Use the three-column chart to organize the review vocabulary in this lesson. Write the word in Spanish. Then write the correct terms to complete each definition.

| English | Spanish | Definition |
|---|---|---|
| **multiplication** | | An operation on two numbers to find their _____. It can be thought of as repeated _____. |
| **denominator** | | The _____ number in a fraction. It represents the number of _____ in the whole. |
| **numerator** | | The _____ number in a fraction; the part of the fraction that tells the number of _____ you have. |
| **fraction** | | A number that represents part of a _____ or part of a _____. |
| **simplest form** | | A _____ in which the greatest common factor of the numerator and the denominator is ____. |

108  Grade 5 • Chapter 10 *Multiply and Divide Fractions*

NAME _____ DATE _____

# Lesson 7 Concept Web

*Multiply Mixed Numbers*

Use the concept web to identify whether mixed numbers are in simplest form. Write *true* or *false*.

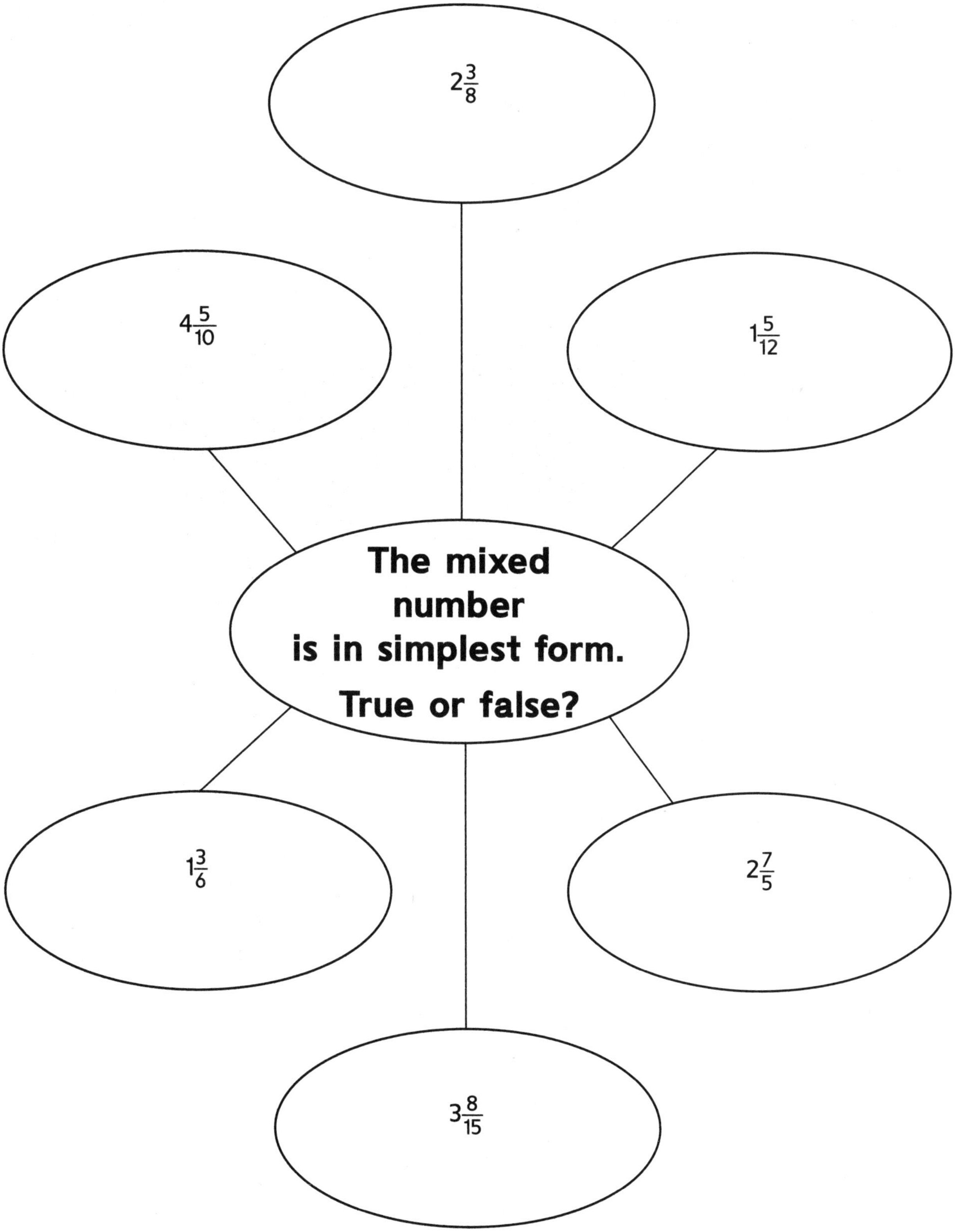

$2\frac{3}{8}$

$4\frac{5}{10}$

$1\frac{5}{12}$

**The mixed number is in simplest form. True or false?**

$1\frac{3}{6}$

$2\frac{7}{5}$

$3\frac{8}{15}$

Grade 5 • Chapter 10 *Multiply and Divide Fractions* **109**

NAME _____ DATE _____

# Lesson 8 Vocabulary Definition Map
*Inquiry/Hands On: Multiplication as Scaling*

Use the definition map to write a description and list characteristics about the vocabulary word or phrase. Write or draw math examples. Share your examples with a classmate.

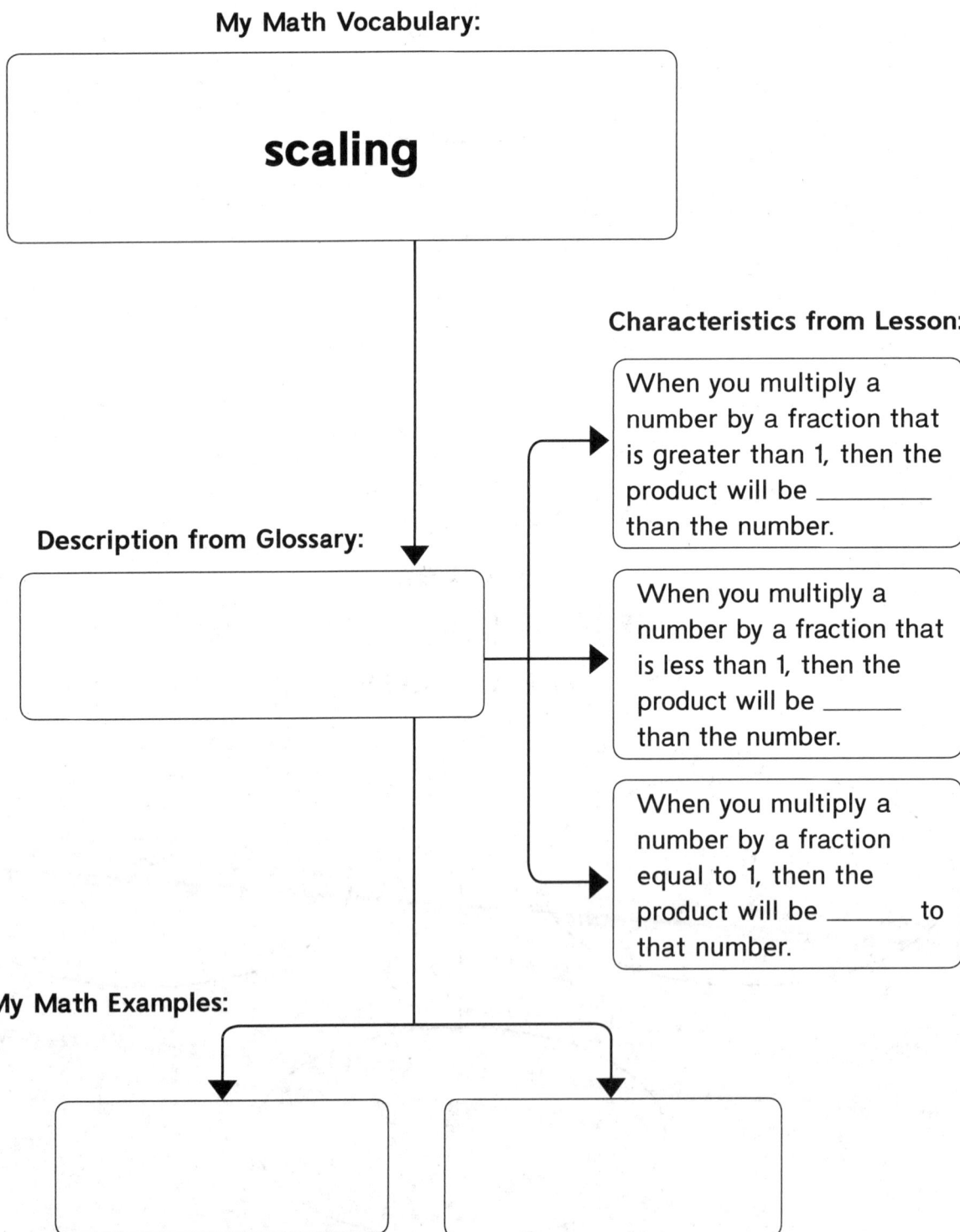

My Math Vocabulary:

**scaling**

**Description from Glossary:**

**Characteristics from Lesson:**

When you multiply a number by a fraction that is greater than 1, then the product will be _____ than the number.

When you multiply a number by a fraction that is less than 1, then the product will be _____ than the number.

When you multiply a number by a fraction equal to 1, then the product will be _____ to that number.

**My Math Examples:**

110 Grade 5 • Chapter 10 *Multiply and Divide Fractions*

NAME _____   DATE _____

# Lesson 9 Guided Writing

*Inquiry/Hands On: Division with Unit Fractions*

**How do you divide with unit fractions?**

Use the exercises below to help you build on answering the Essential Question. Write the correct word or phrase on the lines provided.

1. Rewrite the question in your own words.
   _____
   _____

2. What key words do you see in the question?
   _____

3. A numerator is the _____ number in a fraction. A numerator is the part of the fraction that tells the number of _____ you have.

4. The numerator in the fraction $\frac{1}{5}$ is ____. A unit fraction is a fraction with ____ as its numerator. The fraction $\frac{1}{5}$ is a _____ fraction.

5. When you are dividing $2 \div \frac{1}{5}$ you are trying to find how many _____ of $\frac{1}{5}$ are in ____.

6. How many $\frac{1}{5}$-fraction tiles represent one whole? ____

7. How many $\frac{1}{5}$-fraction tiles represent two wholes? ____

8. $2 \div \frac{1}{5} =$ _____

9. How do you divide with unit fractions?
   _____
   _____

Grade 5 • Chapter 10 *Multiply and Divide Fractions* 111

NAME _____  DATE _____

# Lesson 10 Vocabulary Cognates

*Divide Whole Numbers by Unit Fractions*

Use the Glossary to define the math word in English and in Spanish in the word boxes. Write a sentence using your math word.

| **numerator** | **numerador** |
|---|---|
| Definition | Definición |
| My math word sentence: | |

| **unit fraction** | **fracción unitaria** |
|---|---|
| Definition | Definición |
| My math word sentence: | |

NAME _____  DATE _____

# Lesson 11 Vocabulary Definition Map

*Divide Unit Fractions by Whole Numbers*

Use the definition map to write a description and list characteristics about the vocabulary word or phrase. Write or draw math examples. Share your examples with a classmate.

My Math Vocabulary:

**unit fraction**

**Characteristics from Lesson:**

A _____ is the top number in a fraction. It tells the number of _____ you have.

A _____ is the bottom number in a fraction. It represents the number of parts in the _____.

When you divide a unit fraction by a whole number, the quotient is a _____ _____.

$\frac{1}{2} \div 2 = \frac{1}{4}$; $\frac{1}{4} \div 3 = \frac{1}{12}$; $\frac{1}{5} \div 4 = \frac{1}{20}$

**Description from Glossary:**

**My Math Examples:**

Grade 5 • Chapter 10 *Multiply and Divide Fractions*  113

NAME _____ DATE _____

# Lesson 12 Problem-Solving Investigation

**STRATEGY:** *Draw a Diagram*

**Draw a diagram to solve each problem.**

1. **Mrs. Vallez** purchased sand toys that were **originally** $20. She (Mrs. Vallez) received $\frac{1}{4}$ off of the <u>total</u> price. How much did she **save**?

| Understand | Solve |
|---|---|
| I know: | |
| I need to find: | |
| **Plan** | **Check** |
| ├─────── 20 ───────┤ ☐☐☐☐ | |

2. **Sue** has <u>four</u> DVDs and **Terry** has <u>six</u> DVDs. **They** put all their DVDs <u>together</u> and sold them for **$10** for <u>two</u> DVDs. How **much money** will they earn if they sell **all** of their DVDs?

| Understand | Solve |
|---|---|
| I know: | |
| I need to find: | |
| **Plan** | **Check** |
| ☐☐☐☐☐ ☐☐☐☐☐ | |

114 Grade 5 • Chapter 10 *Multiply and Divide Fractions*

# Chapter 11 Measurement

## Inquiry of the Essential Question:

How can I use measurement conversions to solve real-world problems?

Read the Essential Question. Describe your observations (I see...), inferences (I think...), and prior knowledge (I know...) of each math example. Write additional questions you have below. Then share your ideas and questions with a classmate.

45 ft = _____ yd                                I see...

Since feet are smaller units than yards, divide.

3 feet = 1 yard, so divide by 3.                I think...

45 ÷ 3 = 15

So, 45 feet = 15 yards.                         I know...

---

6 c = _____ fl oz                               I see...

Since cups are larger units than fluid ounces, multiply.

8 fluid ounces = 1 cup, so multiply by 8.       I think...

6 × 8 = 48

So, 6 cups = 48 fluid ounces.                   I know...

---

**Distance Walked (mi)**                        I see...

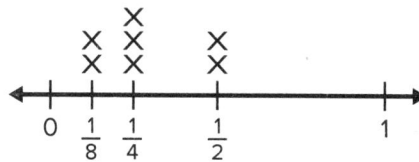
                                                I think...

Fair Share: $\frac{1}{2} + \frac{1}{2} + \frac{1}{4} + \frac{1}{4} + \frac{1}{4} + \frac{1}{8} + \frac{1}{8}$    I know...
$= 2 \div 7 = \frac{2}{7}$ mile

---

Questions I have...

_____

_____

_____

Grade 5 • Chapter 11 Measurement    **115**

# Lesson 1 Note Taking

*Inquiry/Hands On: Measure with a Ruler*

Read the question. Write words you need help with and research each word. Use your lesson to write your Cornell notes. Write or draw math examples to explain your thinking. Share your examples with a classmate.

**Building on the Essential Question**

How do you use a ruler to measure?

**Words I need help with:**

**Notes:**

A ruler can be used to measure _____. Length is the measurement of the _____ between two points.

You can think of a _____ like a number line.

When measuring an object using an inch ruler, it is important to line up the _____ on the ruler with the object.

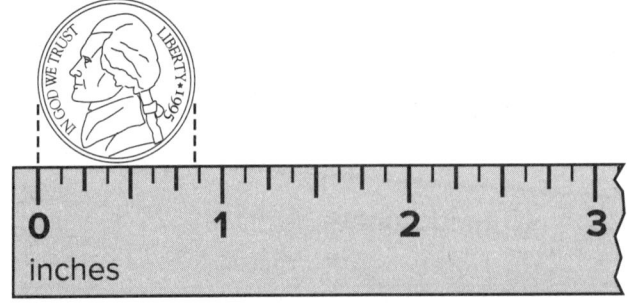

The right side of the nickel is between ___ and ___. The end of the nickel is closest to ___ inch.

If you use a smaller unit of measure, you will get a more _____ measurement.

To the nearest fourth inch, the nickel is **about** $\frac{\_\_}{4}$ inches long.

**My Math Examples:**

NAME _____  DATE _____

# Lesson 2 Vocabulary Chart
## *Convert Customary Units of Length*

Use the three-column chart to organize the vocabulary in this lesson. Write the word in Spanish. Then write the correct terms to complete each definition.

| English | Spanish | Definition |
|---|---|---|
| **convert** | | To change one _____ to another. |
| **customary system** | | The units of _____ most often used in the United States. These include foot, pound, and quart. |
| **foot (ft)** | | A _____ unit for measuring length. Plural is _____. 1 foot = 12 _____ |
| **inch (in.)** | | A _____ unit for measuring length. Plural is _____. |
| **mile (mi)** | | A _____ unit for measuring length. Plural is _____. 1 mile = 5,280 _____ |
| **yard (yd)** | | A _____ unit of length equal to 3 _____ or 36 _____. |

Grade 5 • Chapter 11 *Measurement*  117

NAME _____  DATE _____

# Lesson 3 Problem-Solving Investigation

## STRATEGY: Use Logical Reasoning

Use logical reasoning to solve each problem.

1. An after-school club is building a clubhouse that has a **rectangular** floor that is **8 feet** by **6 feet**.
   What is the **total** floor **area** in <u>square inches</u> of the club?

| Understand | Solve |
|---|---|
| I know the measurements in feet are: | The total area in **square** _____ is: |
| I need to find: | |
| **Plan** | **Check** |
| The measurements in inches are: | |

2. There is a **red**, a **green**, and a **yellow** bulletin board hanging in the hallway.
   <u>All</u> of the bulletin boards are **rectangular** with a **height** of **4** feet.
   Their <u>lengths</u> are <u>6</u> feet, <u>5</u> feet, and <u>3</u> feet.
   The <u>red</u> bulletin board has the <u>largest</u> area and
   the <u>yellow</u> one has the <u>smallest</u> area.
   What is the <u>area</u> of the <u>green</u> bulletin board?

| Understand | Solve |
|---|---|
| I know : | The area of the _____ board is: |
| Largest area = _____ board | |
| Smallest area = _____ board | |
| I need to find: | |
| **Plan** | **Check** |
| The three areas will be: | |
| 4 feet by ___ feet | |
| 4 feet by ___ feet | |
| 4 feet by ___ feet | |

NAME _____  DATE _____

# Lesson 4 Guided Writing

*Inquiry/Hands On: Estimate and Measure Weight*

**How do you estimate and measure weight?**

Use the exercises below to help you build on answering the Essential Question. Write the correct word or phrase on the lines provided.

1. Rewrite the question in your own words.
   _____
   _____

2. What key words do you see in the question?
   _____

3. Weight is a measurement that tells how _____ an object is.

4. You can use a _____ to find the weight of objects. When objects on one side of the balance and objects on the other side are _____, then they are equal in weight.

5. Some of the _____ used to measure weight are ounces (oz) and pounds (lb).

6. Which unit of weight is greater, an ounce or a pound?
   _____

7. There are ___ ounces in one pound.
   There are ___ ounces in two pounds.
   There are ___ ounces in three pounds.

8. A slice of bread weighs about 1 _____ and a loaf of bread weighs about 1 _____.

9. When measuring flour to make one loaf of bread, _____ would give you a more precise measurement.

10. How do you estimate and measure weight?
    _____
    _____
    _____

Grade 5 • Chapter 11 *Measurement*

NAME _____ DATE _____

# Lesson 5 Vocabulary Definition Map
## Convert Customary Units of Weight

Use the definition map to write a description and list characteristics about the vocabulary word or phrase. Write or draw math examples. Share your examples with a classmate.

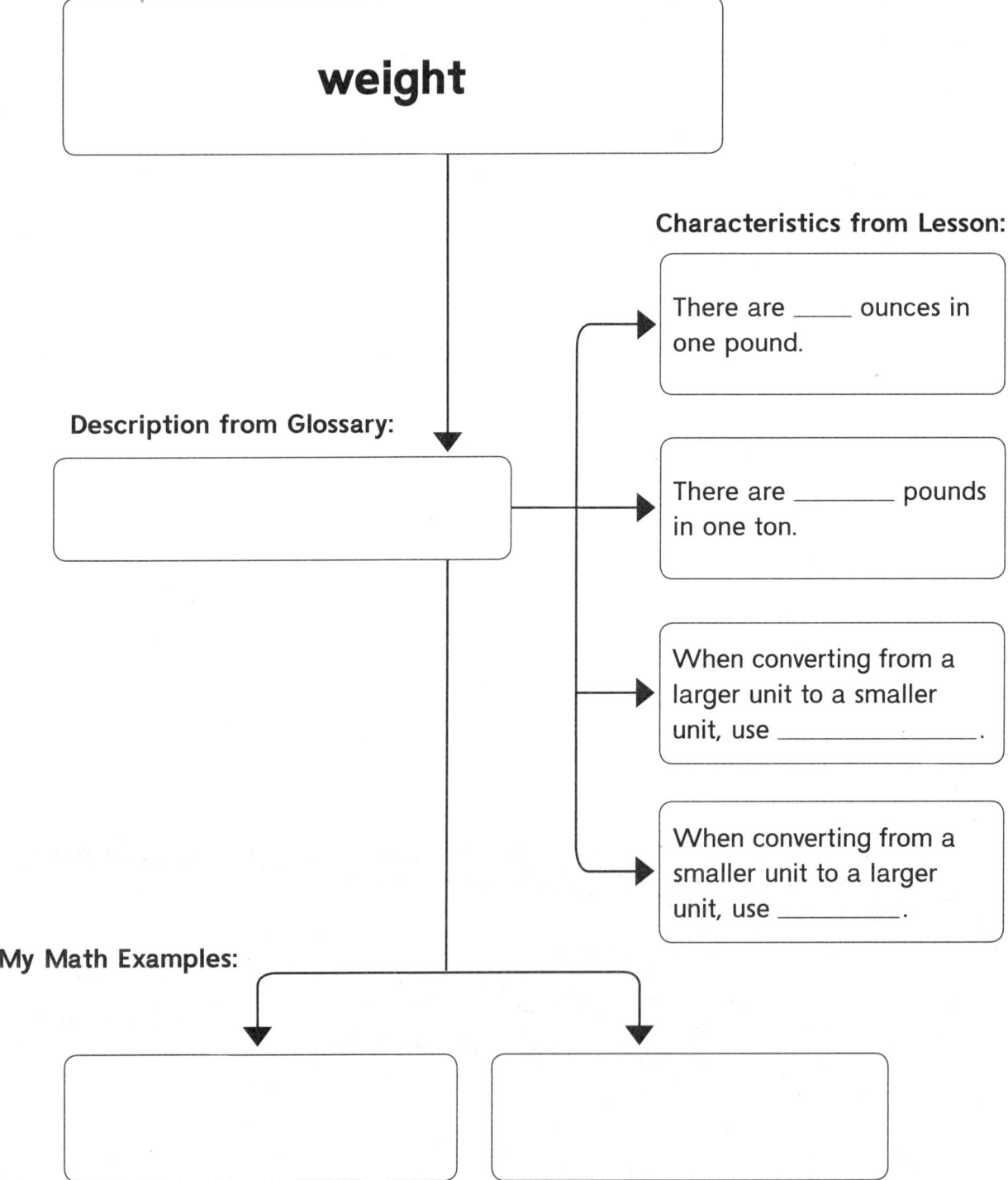

My Math Vocabulary:

**weight**

Description from Glossary:

Characteristics from Lesson:

There are _____ ounces in one pound.

There are _____ pounds in one ton.

When converting from a larger unit to a smaller unit, use _____.

When converting from a smaller unit to a larger unit, use _____.

My Math Examples:

120  Grade 5 • Chapter 11 *Measurement*

NAME _____    DATE _____

# Lesson 6 Note Taking

*Inquiry/Hands On: Estimate and Measure Capacity*

Read the question. Write words you need help with and research each word. Use your lesson to write your Cornell notes. Write or draw math examples to explain your thinking. Share your examples with a classmate.

| Building on the Essential Question | Notes: |
|---|---|
| How do you estimate and measure capacity? | Capacity is a measurement that tells the _____ a container can hold. <br><br> Some of the _____ used to measure capacity are cups (c), pints (pt) quarts (qt), and gallons (gal). <br><br> There are ___ cups in one pint. <br> There are ___ pints in one quart. <br> There are ___ quarts in one gallon.  |
| Words I need help with: | A pint container can hold ___ cups. A container twice that size will be 2 pints in capacity and can hold ___ cups. <br><br> A quart container can hold ___ pints. A container four times that size will be 4 quarts in capacity and can hold ___ pints. |

| Container | Cups | Pints | Quarts |
|---|---|---|---|
| 2 pints |  | 2 | 1 |
| 1 gallon |  |  | 4 |

**My Math Examples:**

Grade 5 • Chapter 11 *Measurement*  121

NAME _____ DATE _____

# Lesson 7 Vocabulary Chart

*Convert Customary Units of Capacity*

Use the three-column chart to organize the vocabulary in this lesson. Write the word in Spanish. Then write the correct terms to complete each definition.

| English | Spanish | Definition |
|---|---|---|
| **capacity** | | The amount a container can _____. |
| **fluid ounce (fl oz)** | | A _____ unit of capacity. |
| **gallon (gal)** | | A customary unit for measuring capacity for _____. 1 gallon = 4 _____ |
| **cup (c)** | | A customary _____ of capacity equal to 8 fluid _____. |
| **pint (pt)** | | A customary unit for _____ capacity. 1 pint = 2 _____ |
| **quart (qt)** | | A customary unit for measuring _____. 1 quart = 4 _____ |

122 Grade 5 • Chapter 11 *Measurement*

NAME _____  DATE _____

# Lesson 8 Vocabulary Definition Map

*Display Measurement Data on a Line Plot*

Use the definition map to write a description and list characteristics about the vocabulary word or phrase. Write or draw math examples. Share your examples with a classmate.

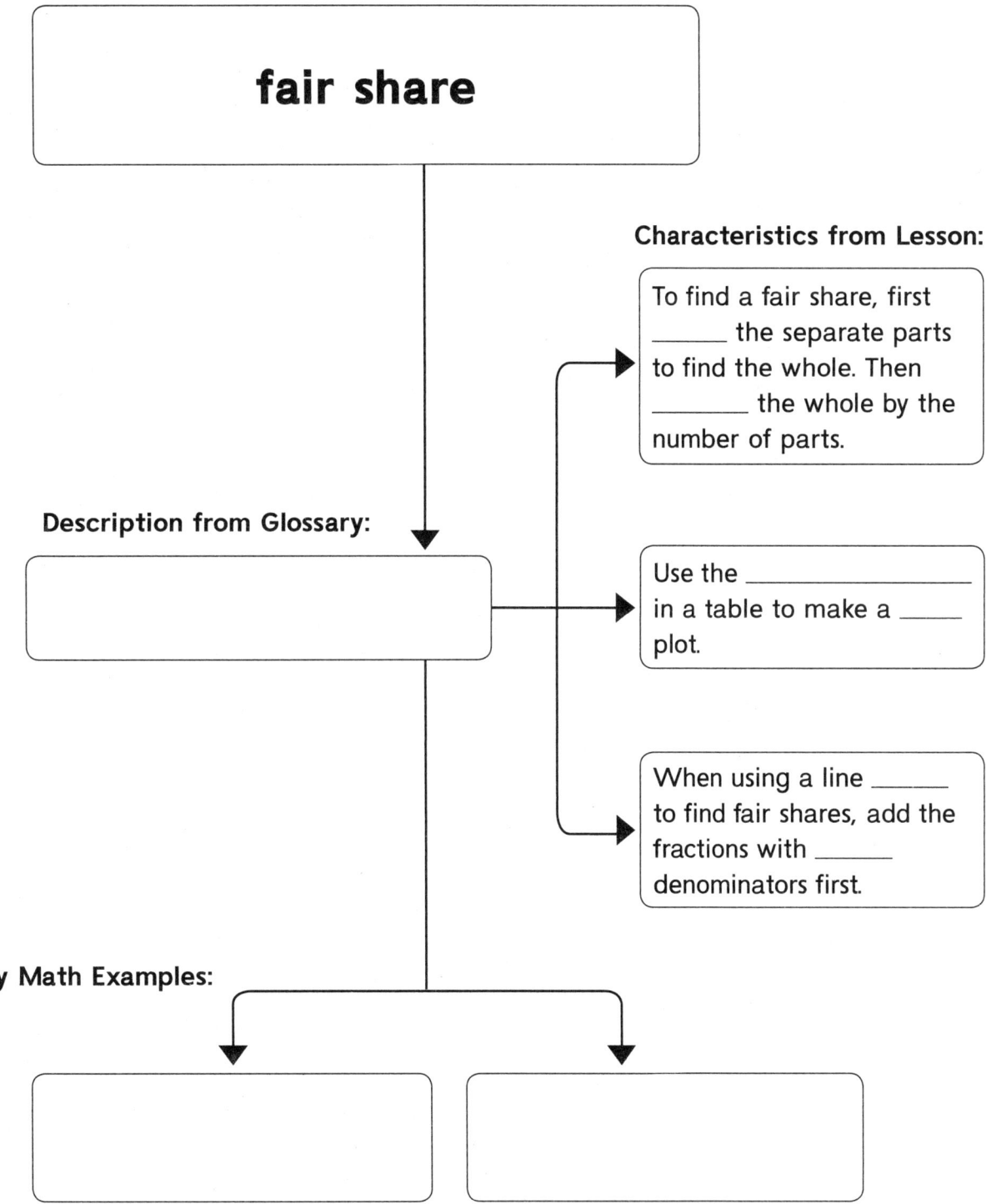

My Math Vocabulary:

**fair share**

**Characteristics from Lesson:**

To find a fair share, first _____ the separate parts to find the whole. Then _____ the whole by the number of parts.

**Description from Glossary:**

Use the _____ in a table to make a _____ plot.

When using a line _____ to find fair shares, add the fractions with _____ denominators first.

**My Math Examples:**

Grade 5 • Chapter 11 *Measurement* **123**

NAME _____ DATE _____

# Lesson 9 Guided Writing

## Inquiry/Hands On: Metric Rulers

**How do you use metric rulers to measure?**

Use the exercises below to help you build on answering the Essential Question. Write the correct word or phrase on the lines provided.

1. Rewrite the question in your own words.
   _____
   _____

2. What key words do you see in the question?
   _____

3. The metric system is a _____ system of measurement.

4. _____ units of length include millimeters, centimeters, meters, and kilometers. Length is the measurement of the _____ between two points.

5. A metric ruler can be used to measure _____ in centimeters and millimeters.
   1 centimeter = ____ millimeters

6. When measuring an object using a metric ruler, it is important to line up the _____ on the ruler with the left side of the object.

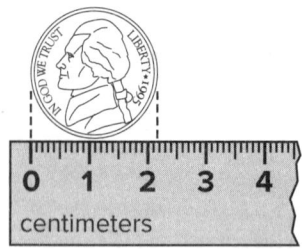

7. The right side of the nickel is between ____ and ____. The end of the nickel is closest to ____ centimeters.

8. If you use a smaller unit of measure, you will get a more _____ measurement.

9. To the nearest millimeter, the nickel is about ____ millimeters long.

10. How do you use metric rulers to measure?
    _____
    _____

124  Grade 5 • Chapter 11 *Measurement*

NAME _____ DATE _____

# Lesson 10 Concept Web

*Convert Metric Units of Length*

Use the concept web to identify the unit of metric measurement that would be best to measure each item.

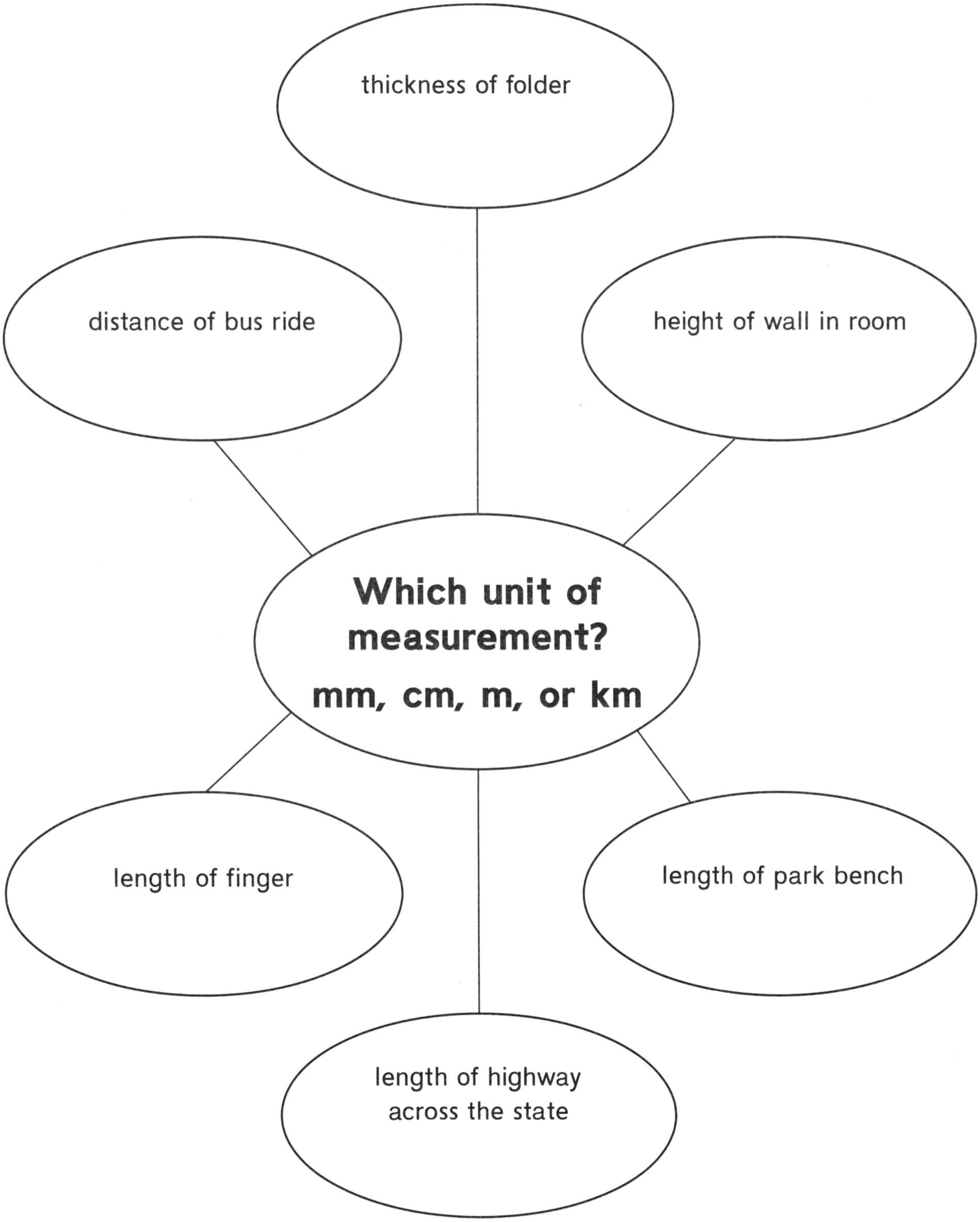

- thickness of folder
- distance of bus ride
- height of wall in room

**Which unit of measurement?**
**mm, cm, m, or km**

- length of finger
- length of park bench
- length of highway across the state

Grade 5 • Chapter 11 *Measurement* **125**

NAME _____ DATE _____

# Lesson 11 Vocabulary Chart

*Inquiry/Hands On: Estimate and Measure Metric Mass*

Use the three-column chart to organize the vocabulary in this lesson. Write the word in Spanish. Then write the correct terms to complete each definition.

| English | Spanish | Definition |
|---|---|---|
| **mass** | | Measure of the amount of _____ in an object. |
| **gram** | | A _____ unit for measuring mass. |
| **kilogram** | | A metric unit for measuring _____. 1 kilogram = 1,000 _____. |
| **metric system** | | The _____ system of measurement. Includes units such as meter, _____, and liter. |

126 Grade 5 • Chapter 11 *Measurement*

NAME _____   DATE _____

# Lesson 12 Multiple Meaning Word
## *Convert Metric Units of Mass*

Complete the four-square chart to review the multiple meaning word.

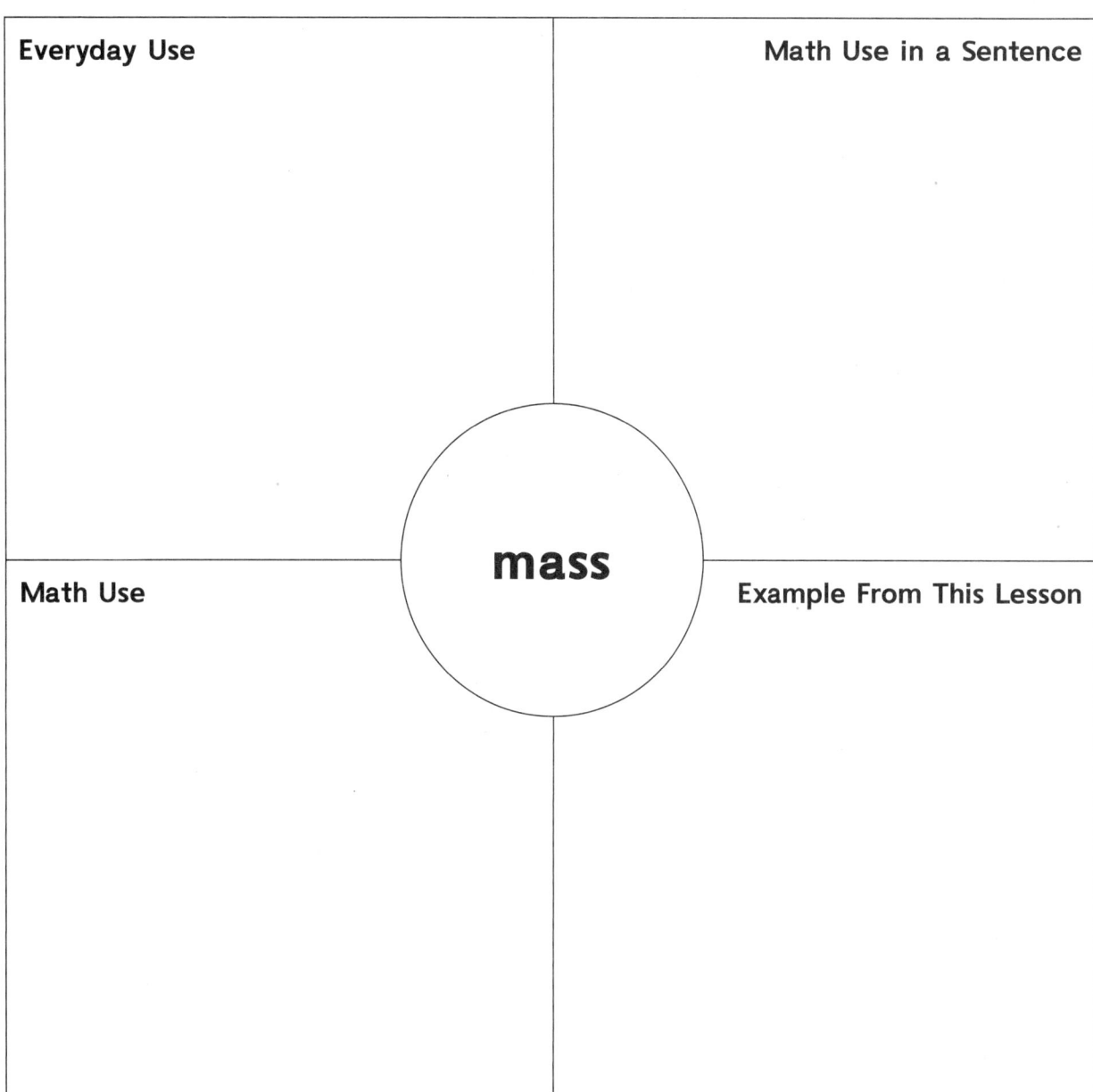

Write the correct term on each line to complete the sentence.

Use _____ to convert from a larger unit of mass to a smaller unit.

Use _____ to convert from a smaller unit of mass to a larger unit.

Grade 5 • Chapter 11 *Measurement*  127

NAME _____ DATE _____

# Lesson 13 Vocabulary Cognates

*Convert Metric Units of Capacity*

Use the Glossary to define the math word in English and in Spanish in the word boxes. Write a sentence using your math word.

| **liter** | **litro (L)** |
|---|---|
| Definition | Definición |
| My math word sentence: ||

| **milliliter** | **mililitro (mL)** |
|---|---|
| Definition | Definición |
| My math word sentence: ||

128 Grade 5 • Chapter 11 *Measurement*

NAME _____  DATE _____

# Chapter 12 Geometry

*Inquiry of the Essential Question:*

**How does geometry help me solve problems in everyday life?**

Read the Essential Question. Describe your observations (I see..), inferences (I think...), and prior knowledge (I know...) of each math example. Write additional questions you have below. Then share your ideas and questions with a classmate.

The figure has 6 sides. The sides are not congruent and the angles are not congruent. So, the polygon is a hexagon that is *not* regular.

I see ...

I think...

I know...

The triangle has no congruent sides and one right angle, So, it is a scalene right triangle.

I see ...

I think...

I know...

The figure has all right angles. Opposite sides are congruent and parallel. So, it is a rectangle.

I see ...

I think...

I know...

Questions I have...

_____

_____

_____

NAME _____ DATE _____

# Lesson 1 Vocabulary Chart
## *Polygons*

Use the three-column chart to organize the vocabulary in this lesson. Write the word in Spanish. Then write the correct terms to complete each definition.

| English | Spanish | Definition |
|---|---|---|
| **congruent angles** | | Angles of a figure that are _____ in measure. This triangle has congruent angles. |
| **congruent sides** | | Sides of a figure that are _____ in length. |
| **hexagon** | | A polygon with _____ sides and _____ angles. |
| **octagon** | | A polygon with _____ sides. |
| **pentagon** | | A polygon with _____ sides. |
| **polygon** | | A _____ figure made up of line segments that do not _____ each other. |
| **regular polygon** | | A polygon in which all **sides** are _____ and all _____ are congruent. |

130  Grade 5 • Chapter 12 *Geometry*

# Lesson 2 Note Taking

*Inquiry/Hands On: Sides and Angles of Triangles*

Read the question. Write words you need help with and research each word. Use your lesson to write your Cornell notes. Write or draw math examples to explain your thinking. Share your examples with a classmate.

| Building on the Essential Question | Notes: |
|---|---|
| How do you describe triangles using sides and angles? | A _____ is a closed figure made up of line segments that do not cross each other. A polygon with three sides and three angles is a _____. 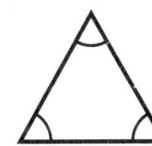 Each angle in this triangle is an _____ angle, which means it measures less than \_\_\_\_ degrees. An angle that measures exactly \_\_\_\_ degrees is a right angle. 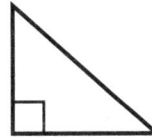 An angle that measures greater than \_\_\_\_ degrees is an obtuse angle. 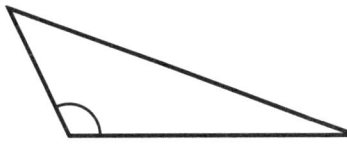 When **angles** have the **same** measure, they are called _____ angles. When **sides** of a triangle measure the **same** length, they are called _____ sides. The _____ of a triangle can be found by adding the side lengths. |
| **Words I need help with:** | |

**My Math Examples:**

Grade 5 • Chapter 12 *Geometry*    **131**

# Lesson 3 Concept Web

## Classify Triangles

Use the concept web to identify each triangle as an isosceles, an equilateral, or a scalene triangle.

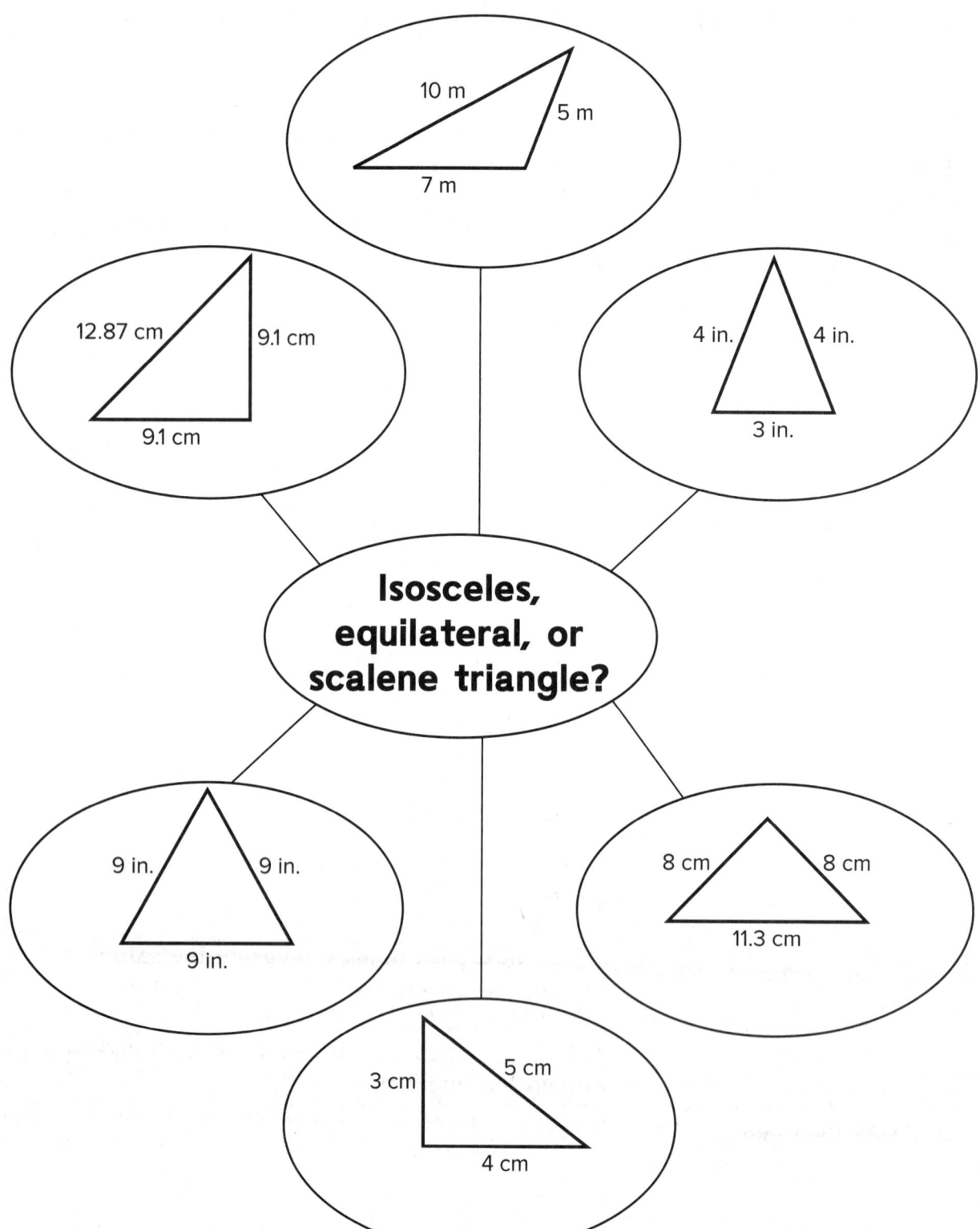

NAME _____ DATE _____

# Lesson 4 Guided Writing

*Inquiry/Hands On: Sides and Angles of Quadrilaterals*

**How do you describe quadrilaterals using sides and angles?**

Use the exercises below to help you build on answering the Essential Question. Write the correct word or phrase on the lines provided.

1. Rewrite the question in your own words.
   _____
   _____

2. What key words do you see in the question?
   _____

3. A _____ is a closed figure made up of line segments that do not cross each other.

4. A polygon that has four sides and four angles is a _____.

5. On each of these quadrilaterals, the arrows point at angles that are across from each other, not next to each other. These angles are called _____ angles.

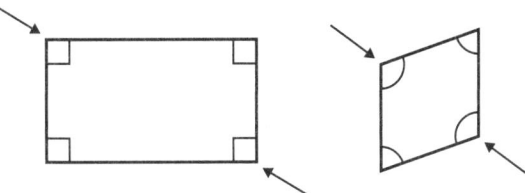

6. Angles that have the same measure are _____ angles.

7. Sides that are the same _____ are congruent sides.

8. Parallel lines are lines that are the _____ distance apart.

9. Opposite sides are sides that are across from each other. The sides do not meet. Opposite sides that are the same distance apart are _____.

10. How do you describe quadrilaterals using sides and angles?
    _____
    _____
    _____

Grade 5 • Chapter 12 *Geometry* **133**

NAME _____ DATE _____

# Lesson 5 Vocabulary Chart
## Classify Quadrilaterals

Use the three-column chart to organize the vocabulary in this lesson. Write the word in Spanish. Then write the correct terms to complete each definition.

| English | Spanish | Definition |
|---|---|---|
| parallelogram | | A quadrilateral with _____ sides in which each pair of opposite sides are _____ and _____. |
| rectangle | | A quadrilateral with _____ right angles; _____ sides are equal and _____. |
| rhombus | | A _____ with _____ congruent sides. |
| trapezoid | | A quadrilateral with exactly _____ pair of _____ sides. |
| square | | A _____ with _____ congruent sides. |

134 Grade 5 • Chapter 12 Geometry

# Lesson 6 Note Taking

*Inquiry/Hands On: Build Three-Dimensional Figures*

Read the question. Write words you need help with and research each word. Use your lesson to write your Cornell notes. Write or draw math examples to explain your thinking. Share your examples with a classmate.

| Building on the Essential Question | Notes: |
|---|---|
| How do I build three-dimensional figures? | A _____-_____ figure is a figure that has length, width, and height. <br><br> A _____ that has rectangular bases is a rectangular prism.  <br><br> A flat surface is called a _____. <br><br> A _____ is a rectangular prism with faces that are congruent squares. 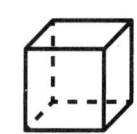 <br><br> Two figures having the same size and the same shape are called _____ figures. <br><br> A two-dimensional pattern of a three-dimensional figure is called a _____. <br><br> The net below is for a rectangular _____, it is made up of _____ rectangles. <br><br> 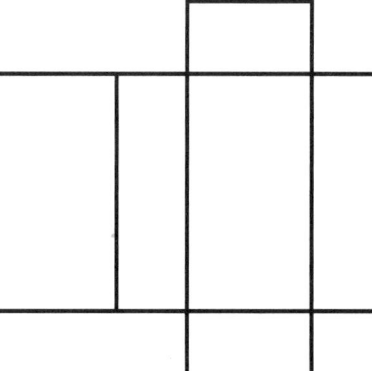 |
| **Words I need help with:** | |
| **My Math Examples:** | |

Grade 5 • Chapter 12 *Geometry*   135

# Lesson 7 Vocabulary Chart
*Three-Dimensional Figures*

Use the three-column chart to organize the vocabulary in this lesson. Write the word in Spanish. Then write the correct terms to complete each definition.

| English | Spanish | Definition |
| --- | --- | --- |
| base | | One of the two parallel _____ faces in a prism. |
| cube | | A rectangular prism with ___ faces that are congruent _____. |
| prism | | A three-dimensional figure with ___ parallel, congruent faces, called _____. At least three faces are _____. |
| rectangular prism | | A prism that has _____ bases. |
| three-dimensional figure | | A figure that has length, width, and _____. |
| triangular prism | | A prism that has _____ bases. |
| vertex | | The point where two _____ meet in an angle or where three or more _____ meet on a three-dimensional figure. |
| edge | | The line segment where two _____ of a three-dimensional figure meet. |
| face | | A _____ surface. |

NAME _____ DATE _____

# Lesson 8 Vocabulary Cognates

*Inquiry/Hands On: Use Models to Find Volume*

Use the Glossary to define the math word in English and in Spanish in the word boxes. Write a sentence using your math word.

| volume | volumen |
|---|---|
| Definition | Definición |
| My math word sentence: | |

| cubic unit | unidad cúbica |
|---|---|
| Definition | Definición |
| My math word sentence: | |

| unit cube | cubo unitario |
|---|---|
| Definition | Definición |
| My math word sentence: | |

Grade 5 • Chapter 12 *Geometry* **137**

NAME _____ DATE _____

# Lesson 9 Multiple Meaning Word
*Volume of Prisms*

Complete the four-square chart to review the multiple meaning word.

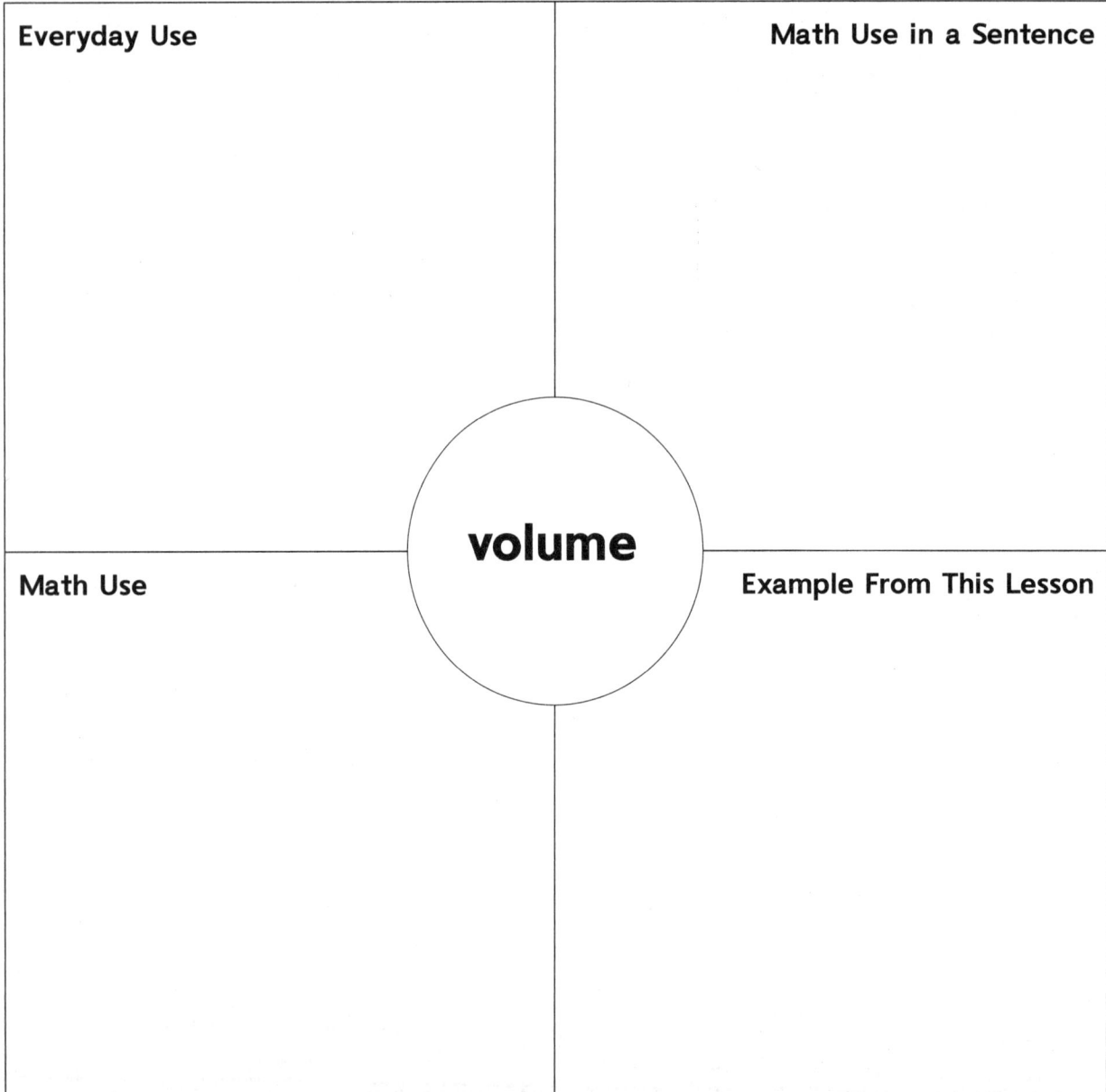

Write the correct term on each line to complete the sentence.

The volume of a container can be found by _____ the length by the _____ by the height.

138  Grade 5 • Chapter 12 *Geometry*

NAME _____ DATE _____

# Lesson 10 Guided Writing

## Inquiry/Hands On: Build Composite Figures

**How do you build composite figures using centimeter cubes?**

Use the exercises below to help you build on answering the Essential Question. Write the correct word or phrase on the lines provided.

1. Rewrite the question in your own words.

   _____

   _____

2. What key words do you see in the question?

   _____

3. _____ is the amount of space inside a three-dimensional figure.

4. A _____ figure is made up of two or more three-dimensional figures.

5. The _____ of a composite figure made with centimeter cubes can be found by _____ the total number of centimeter cubes used to build the composite figure.

6. The volume of the rectangular prism below is ____ cubic centimeters.

   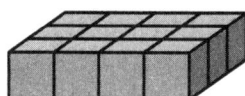

7. The volume of the rectangular prism below is ____ cubic centimeters.

8. The volume of the composite figure below is equal to the _____ of the volume of each rectangular prism above. The volume of this composite figure is ____ cubic centimeters.

9. How do you build composite figures?

   _____

   _____

   _____

   _____

Grade 5 • Chapter 12 Geometry  **139**

NAME _____  DATE _____

# Lesson 11 Vocabulary Definition Map

## Volume of Composite Figures

Use the definition map to write a description and list characteristics about the vocabulary word or phrase. Write or draw math examples. Share your examples with a classmate.

My Math Vocabulary:

**composite figure**

Description from Glossary:

Characteristics from Lesson:

A _____ - _____ figure is a figure that has length, width, and height.

The _____ of a _____ prism is found by multiplying length by width by height.

The volume of a composite figure is equal to the _____ of the volume of each three-dimensional figure that makes up the composite figure.

My Math Examples:

140 Grade 5 • Chapter 12 Geometry

NAME _____ DATE _____

# Lesson 12 Problem-Solving Investigation

## STRATEGY: Make a Model

Solve each problem by making a model.

1. On an assembly line that is **150** feet **long**, there is a work station **every** **15** feet. The **first station** is at the **beginning** of the line. **How many work stations** are there?

| Understand | Solve |
|---|---|
| I know: <br><br> I need to find: | |
| **Plan** <br><br> Label this line to represent the assembly line. <br><br> ●─────────────● | **Check** |

2. A store is stacking cans of food into a **rectangular prism** display.
   The **bottom** **layer** has **8** cans **by** **5** cans.
   There are **5 layers**.
   How **many** cans are in the **display**?

   can

| Understand | Solve |
|---|---|
| I know: <br><br> I need to find: | |
| **Plan** <br> Model the bottom layer. <br><br> How many cans are in the bottom layer? | **Check** |

Grade 5 • Chapter 12 *Geometry* **141**

# What are VKVs® and How Do I Create Them?

Visual Kinethestic Vocabulary Cards® are flashcards that animate words by focusing on their structure, use, and meaning. The VKVs in this book are used to show cognates, or words that are similar in Spanish and English.

## Step 1
Go to the back of your book to find the VKVs for the chapter vocabulary you are currently studying. Follow the cutting and folding instructions at the top of the page. The vocabulary word on the BLUE background is written in English. The Spanish word is on the ORANGE background.

## Step 2
There are exercises for you to complete on the VKVs. When you understand the concept, you can complete each exercise. All exercises are written in English and Spanish. You only need to give the answer once.

## Step 3
Individualize your VKV by writing notes, sketching diagrams, recording examples, and forming plurals.

### How Do I Store My VKVs?
Take a 6" x 9" envelope and cut away a V on one side only. Glue the envelope into the back cover of your book. Your VKVs can be stored in this pocket!

Remember you can use your VKVs ANY time in the school year to review new words in math, and add new information you learn. Why not create your own VKVs for other words you see and share them with others!

# ¿Qué son las VKV y cómo se crean?

Las tarjetas de vocabulario visual y cinético (VKV) contienen palabras con animación que está basada en la estructura, uso y significado de las palabras. Las tarjetas de este libro sirven para mostrar cognados, que son palabras similares en español y en inglés.

## Paso 1
Busca al final del libro las VKV que tienen el vocabulario del capítulo que estás estudiando. Sigue las instrucciones de cortar y doblar que se muestran al principio. La palabra de vocabulario con fondo AZUL está en inglés. La de español tiene fondo NARANJA.

## Paso 2
Hay ejercicios para que completes con las VKV. Cuando entiendas el concepto, puedes completar cada ejercicio. Todos los ejercicios están escritos en inglés y español. Solo tienes que dar la respuesta una vez.

## Paso 3
Da tu toque personal a las VKV escribiendo notas, haciendo diagramas, grabando ejemplos y formando plurales.

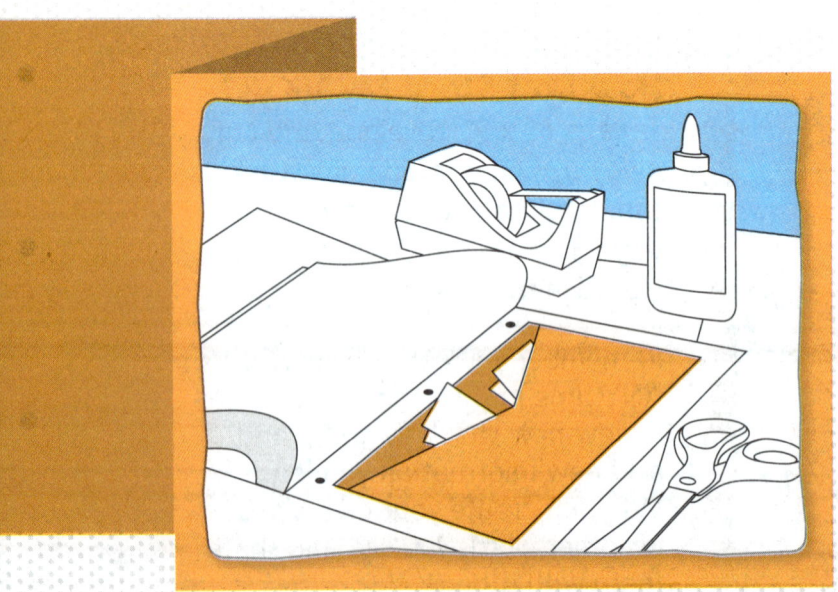

### ¿Cómo guardo mis VKV?
Corta en forma de "V" el lado de un sobre de 6" X 9". Pega el sobre en la contraportada de tu libro. Puedes guardar tus VKV en esos bolsillos. ¡Así de fácil!

Recuerda que puedes usar tus VKV en cualquier momento del año escolar para repasar nuevas palabras de matemáticas, y para añadir la nueva información. También puedes crear más VKV para otras palabras que veas, y poder compartirlas con los demás.

Chapter 1

cut on all dashed lines

fold on all solid lines

**punto**

How is a decimal different from a whole number? (¿En qué se diferencia un decimal de un número entero?)
_____
_____
_____
_____

Write an equivalent decimal for 2.54. (Escribe un decimal equivalente a 2.54.)
_____

**decimal**

**equivalent decimals**

What is another word for *equivalent*? (¿Cuál es otra palabra para decir *equivalente*?)
_____

Chapter 1 Visual Kinesthetic Learning   VKV3

# Chapter 1

cut on all dashed lines

fold on all solid lines

**equivalentes**

Add a decimal point to each equation to make it true. (Agrega un punto decimal a las ecuaciones para que sean verdaderas.)

$\dfrac{3}{100} = 003$

$\dfrac{26}{1,000} = 0026$

$\dfrac{7}{10} = 07$

Write 0.56 in word form. (Escribe 0.56 en palabras.)

Circle each pair of equivalent decimals. (Encierra en un círculo cada par de decimales equivalentes.)

7.25 and 7.52   0.005 and 0.050

0.8 and 0.80   05.2 and 5.20

00.77 and 0.077   3.40 and 3.4

**point**

**decimal**

**decimales**

VKV4 Chapter 1 Visual Kinesthetic Learning

Chapter 2

cut on all dashed lines    fold on all solid lines

Which shows the prime factorization of 12? (¿Cuál muestra la descomposición en factores primos de 12?)

1 × 3 × 4
2 × 2 × 3
1 × 2 × 6

When a number is cubed, it is raised to the _____ power. (Cuando un número es elevado al cubo, es elevado a la _____ potencia.)

Define *exponent*. (Define *exponente*.)

**prime factorization**

**cubed**

Circle the prime numbers. (Encierra en un círculo los números primos.)

2  9
5  8

**exponent**

Chapter 2 Visual Kinesthetic Learning   VKV5

Chapter 2

## ción prima

## al

## e

## o

Use a factor tree to find the prime factorization of 72. (Usa un árbol de factores para hallar la descomposición en factores primos de 72.)

___ × ___ × ___ × ___ = 72

Write each power as a product of the same factor. Then find the value. (Escribe las potencias como el producto del mismo factor. Luego, halla el valor.)

$4^2$ = ___
$10^4$ = ___
$6^3$ = ___

Circle the number that was cubed. (Encierra en un círculo el número que se elevó al cubo.)

Read each description. Write the power and find its value. (Lee las descripciones. Escribe la potencia y halla su valor.)

The base is 4. The exponent is 2. (La base es 4. El exponente es 2.) ___ = ___

The base is 2. The exponent is 3. (La base es 2. El exponente es 3.) ___ = ___

The base is 5. The exponent is 2. (La base es 5. El exponente es 2.) ___ = ___

VKV6  Chapter 2 Visual Kinesthetic Learning

## compatible numbers

**product**

Circle each product. (Encierra en un círculo los productos.)

2 × 3 × 3 = 18

5³ = 75

250 × 4 = 1,000

## estimate

Name two ways you can estimate. (Nombra dos maneras como se puede estimar.)

_____
_____

Compatible numbers are (Los números compatibles son) _____
_____
_____
_____ .

# Chapter 2

**r  o  s**

**números**

Use compatible numbers and mental math to estimate. (Usa números compatibles y cálculo mental para estimar.)

418 × 32 is about (es aproximadamente) _____

97 × 24 is about (es aproximadamente) _____

763 × 47 is about (es aproximadamente) _____

490 × 21 is about (es aproximadamente) _____

Find each product. (Halla los productos.)

3 × 3 × 5 = _____    $4^3$ = _____

$2^4$ = _____    45 × 3 = _____

74 × 5 = _____    $12^2$ = _____

Estimate to find each product. Show how you estimated. (Estima para hallar los productos. Muestra cómo estimaste.)

74 × 59 = _____    488 × 32 = _____

VKV8  Chapter 2 Visual Kinesthetic Learning

Chapter 3

**dividend**

The dividend is (El dividendo es) _____

**quotient**

Circle the quotient. (Encierra en un círculo el cociente.)

$12\overline{)144}$ with 12 on top

Define *quotient*. (Define *cociente*.) _____

**multiple**

Name three multiples of 10. (Nombra tres múltiplos de 10.) _____

Name three multiples of 7. (Nombra tres múltiplos de 7.) _____

Chapter 3 Visual Kinesthetic Learning  VKV9

Chapter 3

**o**     **e**     **o**

**mú**     **coc**

---

An artist finishes 4 paintings each week. How many weeks would it take for the artist to produce 64 paintings? Write a division sentence to solve. Circle the dividend. (Un artista termina 4 pinturas cada semana. ¿Cuántas semanas le tomaría al artista producir 64 pinturas? Escribe una división para resolver. Encierra en un círculo el dividendo.)

_____ ÷ _____ = _____

---

Rewrite each multiplication sentence as a division sentence. Circle the quotients. (Vuelve a escribir cada multiplicación como una división. Encierra en un círculo los cocientes.)

3 × 18 = 54  _____ ÷ _____ = _____

40 × 4 = 160  _____ ÷ _____ = _____

22 × 7 = 154  _____ ÷ _____ = _____

---

Divide mentally. (Divide mentalmente.)

360 ÷ 12 = _____

How did multiples of 10 help you find the answer? (¿Cómo te ayudaron los múltiplos de 10 a hallar la respuesta?)

_____
_____

# Chapter 5

## Associative Property of Addition

Use the Associative Property of Addition to find the sum mentally. Show your steps. (Usa la propiedad asociativa de la suma para hallar mentalmente la suma. Muestra los pasos que tomaste.)

$0.7 + 3 + 5.3 =$ _____

## inverse operations

_____ is the inverse operation of addition. (_____ es la operación inversa de la suma.)

Inverse operations are operations that _____ each other. (Las operaciones inversas son operaciones que _____ entre sí.)

# Chapter 5

*cut on all dashed lines*  
*fold on all solid lines*

**inversas**

**operaciones**

**propiedad asociativa de la suma**

---

Which example shows the Associative Property of Addition? Circle the answer. (¿Cuál ejemplo muestra la propiedad asociativa de la suma? Encierra en un círculo la respuesta.)

1. 3.3 + (5.7 + 4.1) = (3.3 + 5.7) + 4.1
2. 1.8 + 4 = 5.8 and 4 + 1.8 = 5.8
3. 9.6 + 0 = 9.6

---

Solve. Use an inverse operation to check your answer. (Resuelve. Usa una operación inversa para comprobar tu respuesta.)

52.14  
− 2.7

25  
− 12.6

---

Write the correct terms in the sentence. (Escribe las palabras correctas en la oración.)

The Associative Property of Addition states that the way in which numbers are _____ does not change the _____. (La propiedad asociativa en la suma establece que la manera en la que los números están _____ no cambia la _____.)

---

VKV12   Chapter 5 Visual Kinesthetic Learning

# Chapter 6

propiedad conmutativa

## Associative Property of Multiplication

Use the Associative Property of Multiplication to solve mentally. Show your steps. (Usa la propiedad asociativa de la multiplicación para resolver mentalmente. Muestra los pasos que tomaste.)

15 × 9 × 4 = ___

Chapter 6

cut on all dashed lines    fold on all solid lines

propiedad asociativa de la multiplicación

Commutative Property

Which property would you use to find the unknown in the equation below? (¿Cuál propiedad usarías para hallar la incógnita en la siguiente ecuación?)

___ × 2.4 = 2.4 × 9

___ Property of Multiplication

(propiedad ___ de la multiplicación)

VKV14  Chapter 6 Visual Kinesthetic Learning

Chapter 7

## order of operations

Evaluate the expression. (Evalúa la expresión.)

$5^3 + (3 \times 5) - 42 =$ _____

## numerical expression

Is $2^3$ a numerical expression? Explain your answer. (¿Es $2^3$ una expresión numérica? Explica tu respuesta.) _____

Write an example of a numerical expression. (Escribe un ejemplo de una expresión numérica.) _____

## coordinate plane

Circle the origin in the coordinate plane. (Encierra en un círculo el origen en el plano de coordenadas.)

Chapter 7

cut on all dashed lines    fold on all solid lines

**orden de las operaciones**

Write 1–4 to show the correct order of operations. (Escribe los números del 1 al 4 para mostrar el orden correcto de las operaciones.)

___ Add and subtract in order from left to right. (Sumar y restar en orden de izquierda a derecha.)

___ Perform operations in parentheses. (Efectuar las operaciones entre paréntesis.)

___ Find value of exponents. (Hallar el valor de los exponentes.)

___ Multiply and divide in order from left to right. (Multiplicar y dividir en orden de izquierda a derecha.)

**expresión numérica**

Rewrite each numerical expression in a different way. (Vuelve a escribir las expresiones numéricas de otra manera.)

$(3 \times 2) + 5$ _____

$4 + 6 + 6 + 4 + 6$ _____

$(2 \times 7) - (5 \times 2)$ _____

**plano de coordenadas**

Locate and name the ordered pairs in the coordinate plane. (Localiza y representa los pares ordenados en el plano de coordenadas.)

A ( ___ , ___ )
B ( ___ , ___ )
C ( ___ , ___ )
D ( ___ , ___ )

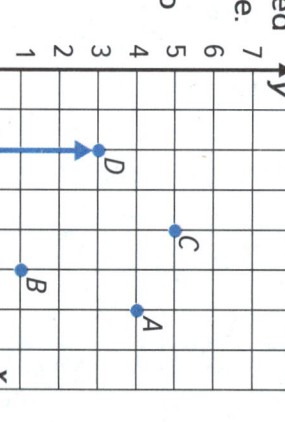

VKV16  Chapter 7 Visual Kinesthetic Learning

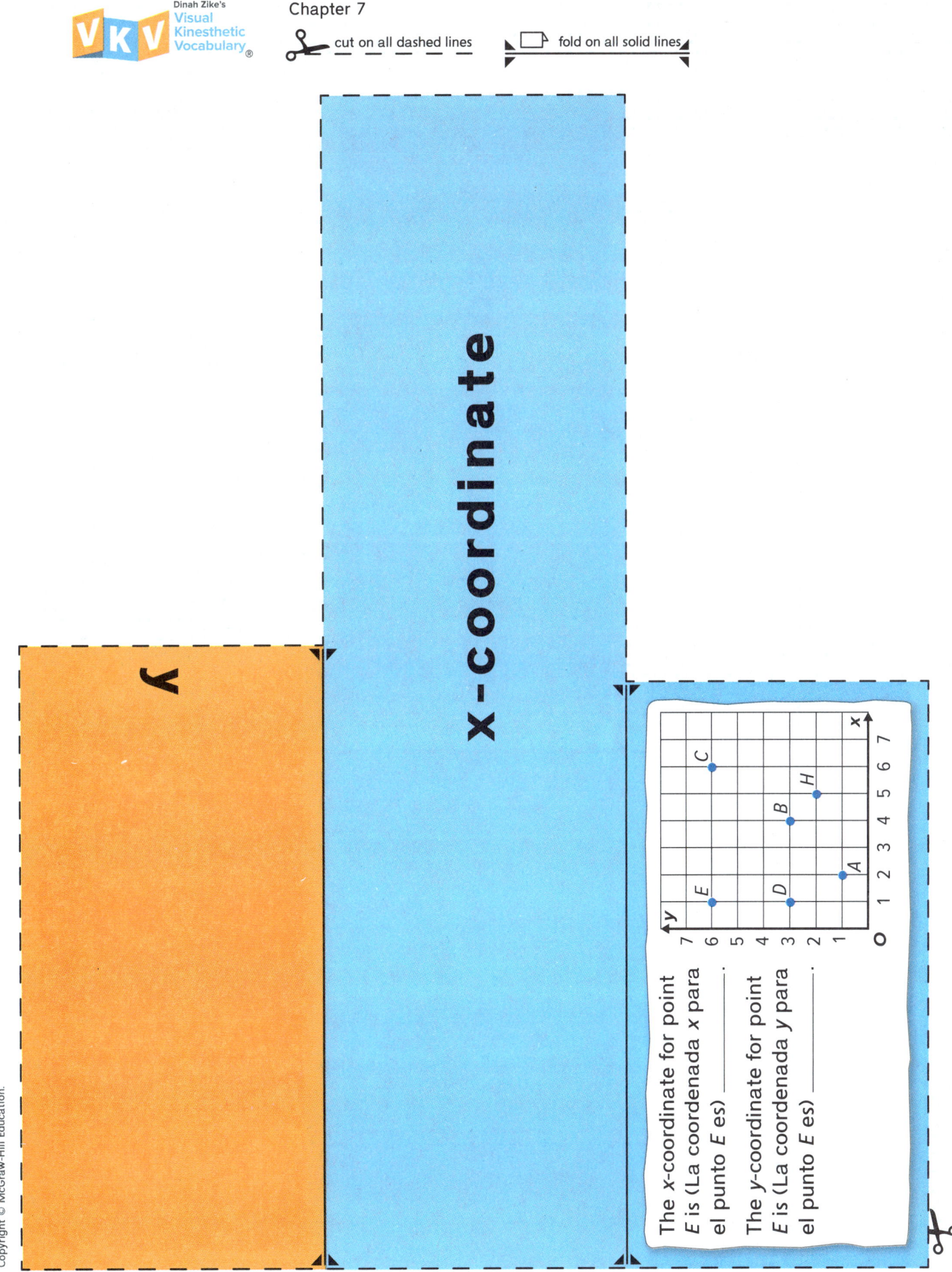

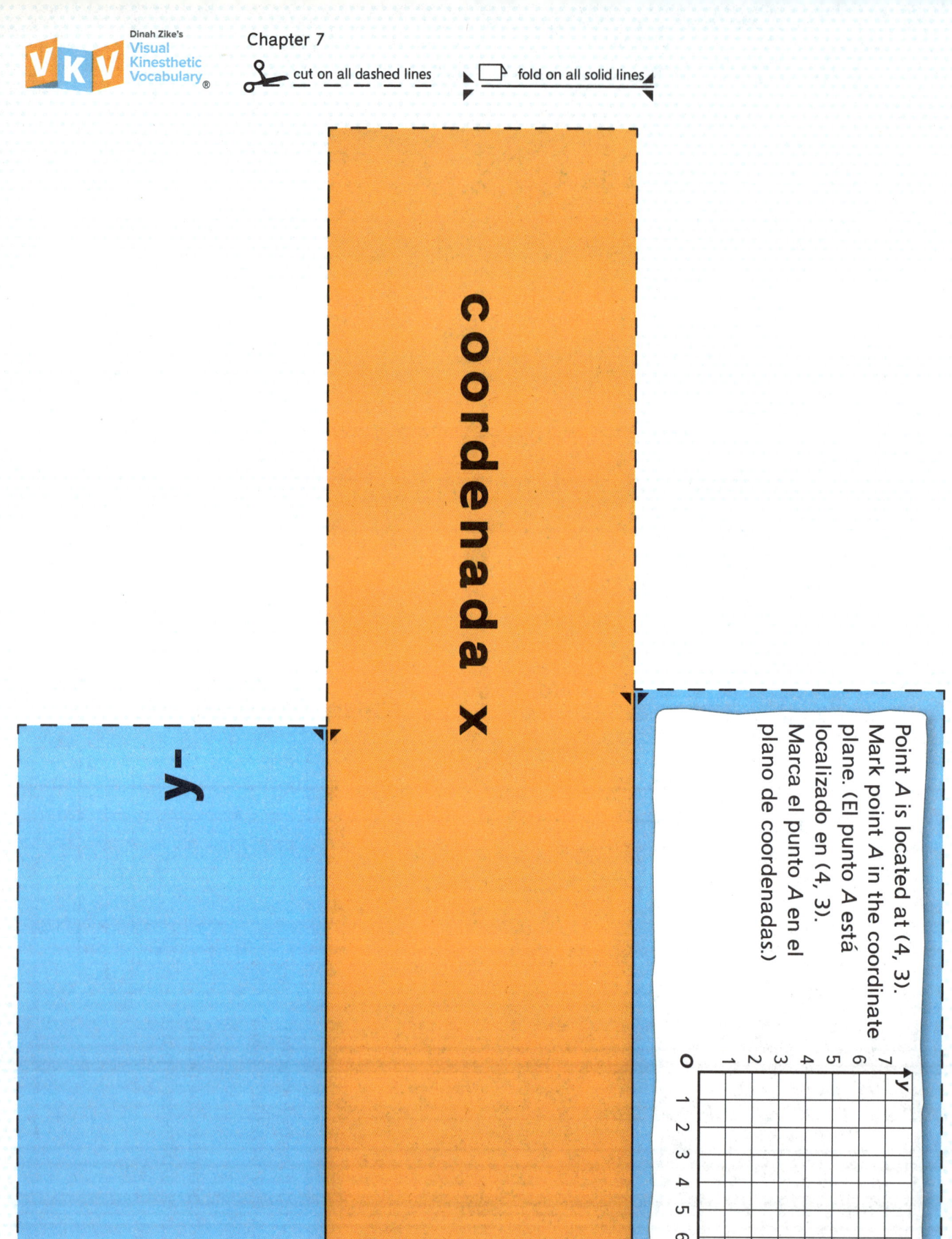

Chapter 8

cut on all dashed lines    fold on all solid lines

What is the first step in finding the GCF for a set of numbers? (¿Cuál es el primer paso para hallar el máximo común divisor de un conjunto de números?)

Circle the fraction that is equivalent to $\frac{7}{21}$. (Encierra en un círculo la fracción que es equivalente a $\frac{7}{21}$.)

$\frac{1}{3}$   $\frac{3}{7}$   $\frac{1}{4}$

**greatest common factor (GCF)**

**equivalent fractions**

Circle the fraction that is equivalent to $\frac{4}{5}$. (Encierra en un círculo la fracción que es equivalente a $\frac{4}{5}$.)

$\frac{8}{15}$   $\frac{2}{3}$   $\frac{20}{25}$

**máximo común divisor (M.C.D.)**

equivalentes

fracciones

Find the GCF for each set of numbers. (Halla el máximo común divisor de cada conjunto de números.)

14, 42, 49

15, 27, 54

Describe how you would simplify the fraction $\frac{18}{63}$. (Describe cómo simplificarías la fracción $\frac{18}{63}$.)

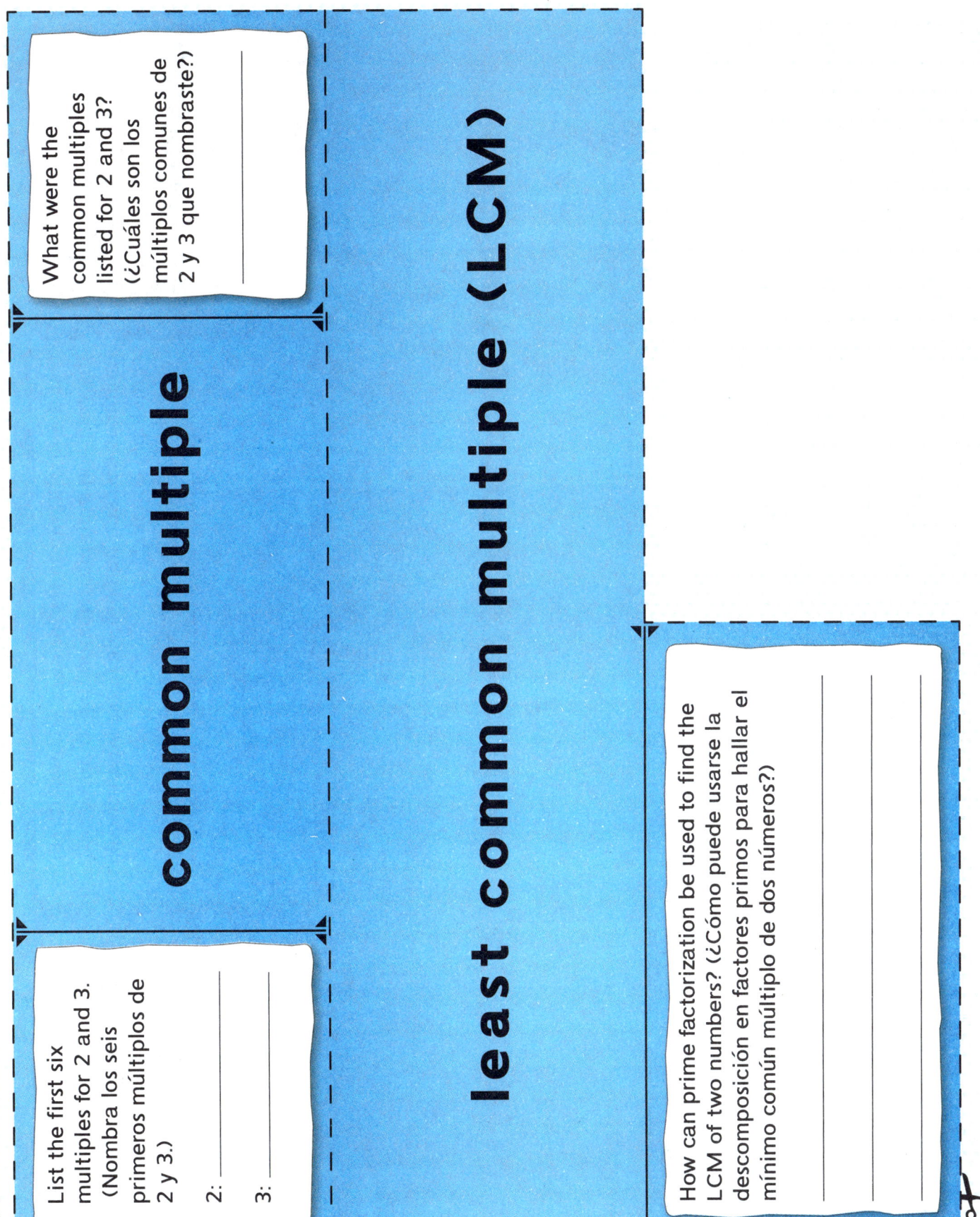

cut on all dashed lines

fold on all solid lines

**común**

**múltiplo**

List the first 10 multiples for each number. Circle the common multiple. (Nombra los 10 primeros múltiplos de cada número. Encierra en un círculo el múltiplo común.)

5: ______________________________

9: ______________________________

**mínimo común múltiplo (M.C.M.)**

Find the LCM of each set of numbers. (Halla el mínimo común múltiplo de cada conjunto de números.)

4, 5, 8 ________

8, 9, 10 ________

Chapter 11

cut on all dashed lines    fold on all solid lines

Circle the word with the same meaning as *convert*. (Encierra en un círculo la palabra que significa lo mismo que *expresar de otra manera*.)

multiply (multiplicar)    measure (medir)

change (convertir)    divide (dividir)

Define *capacity*. (Define *capacidad*.)

Name something that is measured in fluid ounces. (Nombra algo que se mida en onzas líquidas.)

**convert**

**capacity**

**fluid ounce**

How is a fluid ounce different from an ounce? (¿En qué se diferencia una onza líquida de una onza?)

## Chapter 11

**ir**

**dad**

**líquida**

**onza**

---

Complete the equations to convert each measurement. (Completa las ecuaciones para convertir cada medida.)

24 feet (pies) = _____ yards (yardas)

3,520 yards (yardas) = _____ miles (millas)

3 yards (yardas) = _____ inches (pulgadas)

1.5 feet (pies) = _____ inches (pulgadas)

**Customary Units of Length**

1 ft = 12 in.
1 yd = 3 ft or 36 in.
1 mi = 5,280 ft or 1,760 yd

---

Use <, >, or = to make each statement true. (Usa <, > o = para hacer que cada enunciado sea verdadero.)

3 pints (pintas) ◯ 1.5 gallons (galones)

1 quart (cuarto de galón) ◯ 3 cups (tazas)

4 cups (tazas) ◯ 1 quart (cuarto de galón)

**Customary Units of Capacity**

1 cup (taza)(c) = 8 fluid ounces (fl oz)
1 pint (pinta)(pt) = 2 c = 16 fl oz
1 quart (cuarto de galón) (qt) = 2 pt = 32 fl oz
1 gallon (galón) (gal) = 4 qt = 128 fl oz

---

Draw a line from each unit of capacity to its equivalent measurement in fluid ounces. (Traza una línea de cada unidad de capacidad a su medida equivalente en onzas líquidas.)

1 gallon (galón)            8 fl oz (onzas líquidas)

1 cup (taza)                16 fl oz (onzas líquidas)

1 quart (cuarto de galón)   32 fl oz (onzas líquidas)

1 pint (pinta)              128 fl oz (onzas líquidas)

# Chapter 11

cut on all dashed lines

fold on all solid lines

Name three things you might measure in kilometers. (Nombra tres cosas que podrías medir en kilómetros.) ___

There are ___ millimeters in 1 centimeter and ___ centimeters in 1 meter. (Hay ___ milímetros en 1 centímetro y ___ centímetros en 1 metro.)

The metric system is a ___ system of measurement. (El sistema métrico es un sistema de medición ___.)

**kilometer**

**centimeter**

**metric system**

Name two metric units of measurement. (Nombra dos unidades métricas de medida.) ___

# Chapter 11

**métrico**

**ímetro**

**ómetro**

**sistema**

1 kilometer (kilómetro) (km) = 1,000 meters (metros) (m)

Complete each equation. (Completa las ecuaciones.)

2 km = _____ m

1,500 m = _____ km

12 km = _____ m

5.6 km = _____ m

850 m = _____ km

3.54 km = _____ m

Complete each equation. (Completa las ecuaciones.)

2 cm = _____ mm

54 mm = _____ cm

16 cm = _____ mm

1.2 m = _____ cm

254 cm = _____ m

0.6 m = _____ cm

Is it easier to convert between customary units of measurement or metric units of measurement? Explain your answer. (¿Es más fácil convertir entre unidades de medida del sistema usual o unidades métricas de medida? Explica tu respuesta.)

Chapter 11

✂ cut on all dashed lines   📄 fold on all solid lines

**gram**

A _____ has a mass of about 1 gram.
(_____ tiene una masa de aproximadamente 1 gramo.)

A _____ has a mass of about 1 kilogram.
(_____ tiene una masa de aproximadamente 1 kilogramo.)

A liter is a _____ unit of capacity. There are _____ milliliters in 1 liter.
(Un litro es una unidad _____ de capacidad. Hay _____ mililitros en 1 litro.)

**liter**

Grams and kilograms are metric units of _____.
(Los gramos y los kilogramos son unidades métricas de _____.)

**Chapter 11 Visual Kinesthetic Learning   VKV27**

Chapter 11

**kilo gramo**

**kilo**

1 kilogram (kilogramo) (kg) = 1,000 grams (gramos) (g)

Use <, >, or = to make each statement true. (Usa <, > o = para hacer que cada enunciado sea verdadero.)

1.75 kg ◯ 17,500 g
350 g ◯ 0.35 kg
12 kg ◯ 1,200 g

Complete each equation. (Completa las ecuaciones.)

2 L = _____ mL
450 mL = _____ L
6.75 L = _____ mL

3.2 L = _____ mL
5,800 mL = _____ L
0.9 L = _____ mL

Chapter 12

cut on all dashed lines    fold on all solid lines

Congruent angles have the same _____. (Los ángulos congruentes tienen la misma _____.)

A regular octagon has 8 _____ and 8 _____. (Un octágono regular tiene 8 _____ y 8 _____.)

List three polygons. Draw one example. (Nombra tres polígonos. Dibuja un ejemplo.)

**congruent angles**

**hexagon**

**polygon**

Write a word with the same meaning as *congruent*. (Escribe una palabra que signifique lo mismo que *congruente*.)

A regular hexagon has 6 _____ and 6 _____. (Un hexágono regular tiene 6 _____ y 6 _____.)

Chapter 12 Visual Kinesthetic Learning  VKV29

# Chapter 12

cut on all dashed lines  fold on all solid lines

**ígono**

**ágono**

**congruentes**

Circle the figure with congruent angles. (Encierra en un círculo la figura que tiene ángulos congruentes.)

Draw a hexagon and an octagon that are not regular. (Dibuja un hexágono y un octágono que no sean regulares.)

Draw an example of a regular polygon. Why is a circle NOT a polygon? (Dibuja un ejemplo de un polígono regular. ¿Por qué un círculo NO es un polígono?)

**oct**

**ángulos**

VKV30  Chapter 12 Visual Kinesthetic Learning

Chapter 12

cut on all dashed lines   fold on all solid lines

triangle

equilátero

isósceles

Is an isosceles triangle a regular polygon? Explain your answer. (¿Es un triángulo isósceles un polígono regular? Explica tu respuesta.)

Chapter 12 Visual Kinesthetic Learning   VKV31

Chapter 12

cut on all dashed lines

fold on all solid lines

triángulo

isosceles

equilateral

Label each figure as an isosceles triangle or an equilateral triangle. (Rotula las figuras como triángulo isósceles o triángulo equilátero.)

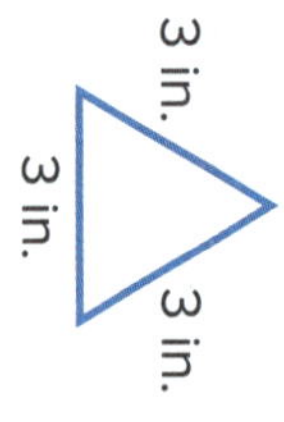

3 in. / 3 in. / 3 in.

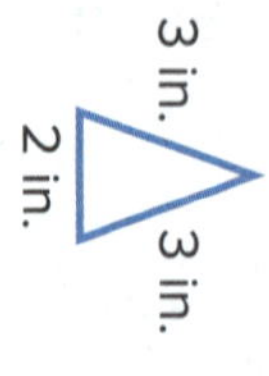

3 in. / 3 in. / 2 in.

**VKV32** Chapter 12 Visual Kinesthetic Learning

Chapter 12

cut on all dashed lines   fold on all solid lines

Draw an example of two congruent figures. (Dibuja un ejemplo de dos figuras congruentes.)

At least three faces of a prism are (Al menos tres caras de un prisma son) _____.

What is the volume of a prism 2 units long, 3 units wide, and 2 units high? (¿Cuál es el volumen de un prisma de 2 unidades de largo, 3 unidades de ancho y 2 unidades de altura?)

____ cubic units
(unidades cúbicas)

**congruent figures**

**prism**

**cubic unit**

Define *congruent figures*. (Define *figuras congruentes*.)

A cubic unit is used for (Una unidad cúbica se usa para) _____.

Chapter 12 Visual Kinesthetic Learning   VKV33

# Chapter 12

**cúbica**

**a**

**congruentes**

**unidad**

**figuras**

_____ cubic units (unidades cúbicas)

What is the volume of the figure below? (¿Cuál es el volumen de la siguiente figura?)

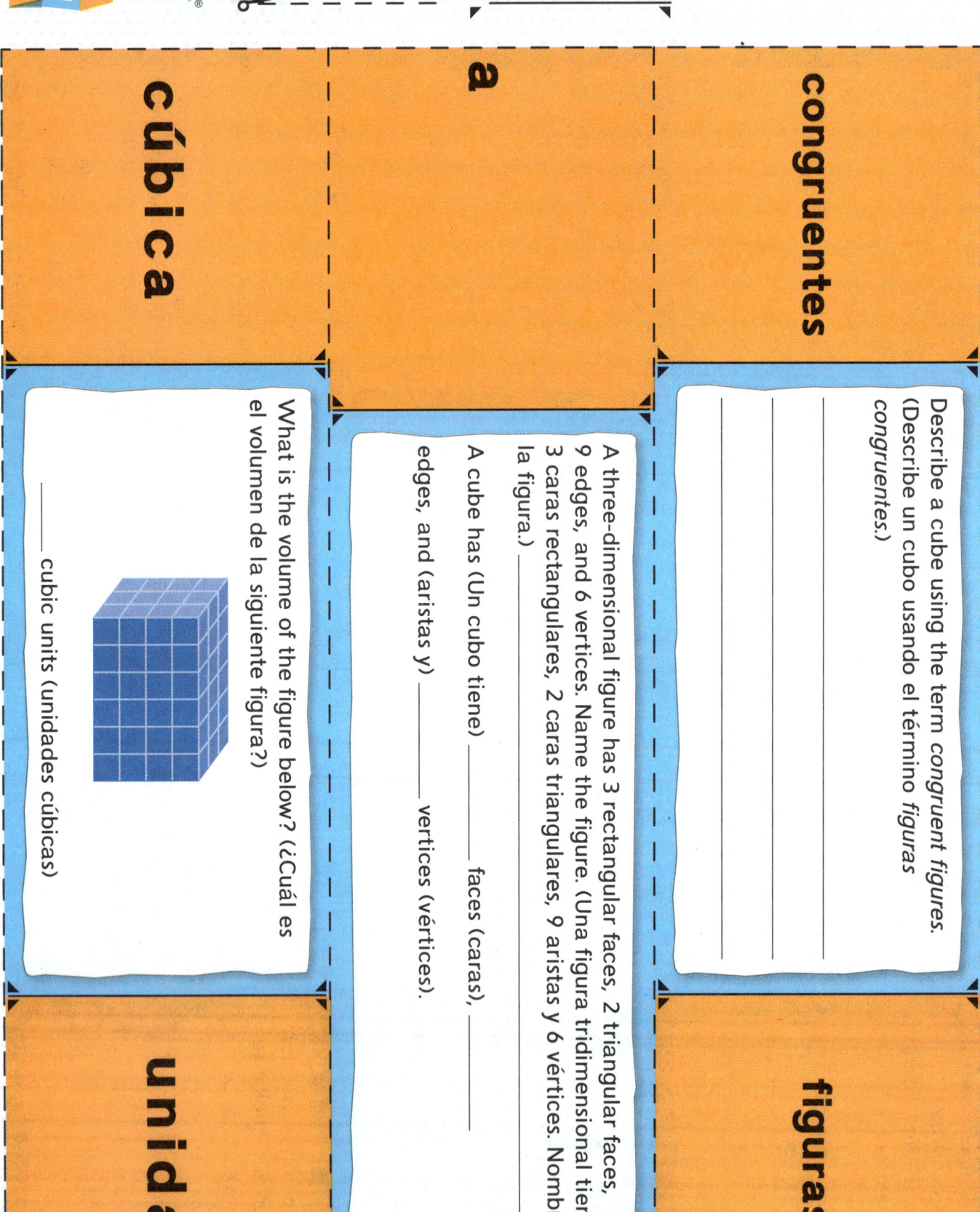

A three-dimensional figure has 3 rectangular faces, 2 triangular faces, 9 edges, and 6 vertices. Name the figure. (Una figura tridimensional tiene 3 caras rectangulares, 2 caras triangulares, 9 aristas y 6 vértices. Nombra la figura.)

A cube has (Un cubo tiene) _____ faces (caras), _____ edges, and (aristas y) _____ vertices (vértices).

Describe a cube using the term *congruent figures*. (Describe un cubo usando el término *figuras congruentes*.)